Selected

Volume

12 **V. S. Varadarajan, Editor**
Algebra in ancient and modern times
1998

11 **Kunihiko Kodaira, Editor**
Basic analysis: Japanese grade 11
1996

10 **Kunihiko Kodaira, Editor**
Algebra and geometry: Japanese grade 11
1996

9 **Kunihiko Kodaira, Editor**
Mathematics 2: Japanese grade 11
1997

8 **Kunihiko Kodaira, Editor**
Mathematics 1: Japanese grade 10
1996

7 **Dmitry Fomin, Sergey Genkin, and Ilia Itenberg**
Mathematical circles (Russian experience)
1996

6 **David W. Farmer and Theodore B. Stanford**
Knots and surfaces: A guide to discovering mathematics
1996

5 **David W. Farmer**
Groups and symmetry: A guide to discovering mathematics
1996

4 **V. V. Prasolov**
Intuitive topology
1995

3 **L. E. Sadovskiĭ and A. L. Sadovskiĭ**
Mathematics and sports
1993

2 **Yu. A. Shashkin**
Fixed points
1991

1 **V. M. Tikhomirov**
Stories about maxima and minima
1990

ALGEBRA IN ANCIENT AND MODERN TIMES

MATHEMATICAL WORLD · VOLUME 12

ALGEBRA
IN ANCIENT AND
MODERN TIMES

V. S. VARADARAJAN

AMERICAN MATHEMATICAL SOCIETY

HINDUSTAN BOOK AGENCY

1991 *Mathematics Subject Classification.* Primary 01–01, 12–01, 12–03;
Secondary 01A20, 01A29, 01A30, 01A40.

ABSTRACT. The book presents the basics of modern algebra at a very elementary level. The author uses a historical approach showing the growth of algebra from its ancient origins to its current state. The material is presented in a way that allows the student to see how the ideas emerged in the attempt to answer specific questions.

Library of Congress Cataloging-in-Publication Data

Varadarajan, V. S.
    Algebra in ancient and modern times / V. S. Varadarajan.
       p. cm. — (Mathematical world, ISSN 1055-9426 ; v. 12)
    ISBN 0-8218-0989-X (alk. paper)
    1. Algebra—History.   I. Title.   II. Series.
QA151.V37   1998
512′.009—dc21
                                                98-15355
                                                  CIP

To my students

CONTENTS

PREFACE

This book arose out of a course entitled

Mathematics in ancient and modern times:

The birth of modern algebra

which I gave to a small group of freshmen students in UCLA in the Fall of 1993. The students had only high school mathematics as their background and so I did not use calculus. However I assisted them so that they selected individual projects to work on as their assignment for fulfilling the requirements of the course. Each project was designed to have two parts: a historical and a mathematical one. The projects had wide variety which surprised me, but which increased my pleasure as the course progressed. This was the first time that I had taught a small group of freshmen in a format different from the usual cook-book type in more than thirty years of teaching in UCLA, and the experience was memorable.

My inspiration for designing and teaching such a course came after reading the beautiful book *Tales of physicists and mathematicians* by my friend Simon Gindikin, especially his account of the life and work of Cardano and its influence on the rise of modern algebra. However in the course I did not confine myself to the work of Cardano but explored a wider theme, namely the growth of algebra from its ancient origins to its current state that allows us to have a marvellous understanding of the whole gamut of physical sciences: from the abstract reaches of number theory to the wonders of quantum physics.

In 1996 Rajendra Bhatia of the Indian Statistical Institute suggested the possibility of making a book out of these notes. I then revised the notes adding exercises and additional material—partly historical and partly dealing with more advanced topics— in the form of notes and exercises at the end of the individual chapters. These may also be used as starting points for projects for the more ambitious student. The result is what I hope is an interesting way to present the basics of modern algebra at a very elementary level. The approach that I have adopted here (as in the course) is what Shafarevich calls the biogenetic one; the student is exposed to the historical and evolutionary development of the subject as an essential part of the course, and the material is presented in a way that allows him or her to see clearly how ideas have emerged in the attempt to solve specific questions.

I am not sure how successful my attempt has been. However it was a great experience for me, not only at the time the course was given, but also later when the book was written. I would like to record my gratefulness to the Honors Collegium of UCLA which encouraged me to give the course, to the small group of young men and women who took the original course, and to R. Bhatia who encouraged me to enlarge the notes into its present form. Finally, I would like to thank Wissam Barakat for his great technical help in the preparation of the final version of the manuscript.

October 1997

V. S. Varadarajan
Department of Mathematics
University of California
Los Angeles, CA 90095–1555
USA

MATHEMATICIANS OF ANCIENT TIMES

- PYTHAGORAS (c. 500 BC)

- EUDOXUS (c. 400 BC)

- ARCHIMEDES (c. 287 BC—212 BC)

- EUCLID (c. 300 BC)

- APOLLONIUS (c. 262 BC—190 BC)

- DIOPHANTUS (c. 250)

- LIU HUI (c. 300)

- ĀRYABHAṬA (476—c. 550)

- BRAHMAGUPTA (c. 598—665)

- AL–KHWARIZMI (c. 780—c. 850)

- BHĀSKARA (c.1114—1185)

MATHEMATICIANS INVOLVED IN THE
BIRTH OF MODERN ALGEBRA

- FIBONACCI (1180—1240)

- FRA LUCA PACIOLI (1445—1514)

- SCIPIONE DEL FERRO (1465—1526)

- NICCOLO TARTAGLIA (c. 1500—1557)

- GEROLAMO CARDANO (1501—1576)

- LUIGI FERRARI (1522—1565)

- RAFFAEL BOMBELLI (1526—1573)

- FRANCOIS VIETE (1540—1603)

SOME HISTORY OF EARLY MATHEMATICS

1

EUCLID–ARCHIMEDES–DIOPHANTUS

In the front of the book there are two tablets containing the names of some mathematicians and their life spans. The first one consists of some mathematicians of ancient times, from Central Asia, China, Greece, and India. The second contains some of the principal characters in the creation of modern algebra, mostly from Italy.

What is most striking about these mathematicians is the originality of their ideas and the fact that after all these centuries these ideas are still alive and inspiring modern research. In the development of mathematics there is therefore a continuity that goes back all the way to the pre-Christian era. It is therefore a good idea to start by looking briefly into some of the ideas and problems that occupied mathematicians from these epochs and try to draw a line from them to what is of interest to us in the present.

EUCLID

Euclid is one of the most famous mathematicians of all time. Almost every educated person knows about him and his work in geometry. Not much is known about his life except that he lived in Alexandria, Greece around 300 BC and had a school of disciples. His achievement, monumental by the standards of any epoch, was to establish Geometry as a deductive science based on a small number of fundamental principles called *axioms*. He did this in a work called the *Elements*, consisting of 13 Books [H–E], which systematically developed the geometrical facts about triangles, circles, and other planar and spatial figures. It is not an exaggeration to say that no mathematical treatise has been more influential than Euclid's *Elements*. Among the most famous of the theorems in the Elements are the following.

The sum of the three angles of a triangle is equal to two right angles.

The area of the square on the hypotenuse of a right angled triangle is equal to the sum of the areas of the squares on the other two sides.

It took nearly two thousand years of effort by mathematicians before a full understanding of Geometry could be reached, and this required the development of *Non–Euclidean geometry* by **Gauss, Bolyai, Lobachevsky** and the completion of their work by **Klein and Beltrami**, the developments on the foundations of mathematics by **Hilbert** and his definitive reformulation of the axiomatic structure of both Euclidean and Non–Euclidean Geometries, and the construction of what are now called *Riemannian Geometries* by **Riemann**. In the words of **Hermann Weyl**, one of the greatest mathematicians of this century,

> If we now look at his (Euclid's) Geometry, it is as if we see a face which is very old and very familiar, but which is sublimely transfigured.

In spite of the fact that our current view of Euclidean Geometry as an axiomatic discipline is so much more subtle than Euclid's, his work still compels our admiration and veneration. He also understood very well that mathematics has to be pursued following its internal esthetic, as illustrated by the comment he made to his slave, in response to a man asking him what is the use of his geometrical propositions.

> Give this man three pence since he must needs make gain by what he learns.

In these days when there is much clamour that only those parts of science that are "applicable to real problems" deserve support it will be well to heed the words of one of the wisest men who ever lived.

Euclid's volumes also included many results on whole numbers, i.e., integers. Everyone has some knowledge of divisibility properties of positive integers. For instance 121 is divisible by 11–in fact $121 = 11 \times 11$ but 11 is not divisible by any number other than 1 and itself. Numbers that possess this last mentioned property are known as *prime numbers*. Thus

$$2, 3, 5, 7, 11, 13, 17, 19, 21, 23, 29, 31, 37, 41, 43, 47, 53, 59, 61, 67, 71, 73, 79,$$

are primes, as one can easily verify. It is then natural to wonder if this sequence keeps going without stopping, namely, whether the sequence of primes is *infinite*. In Euclid one will find the famous proof that this is indeed so:

> The sequence of primes is infinite.

The proof is very simple and yet remarkably beautiful. In the exercises below we shall sketch the argument given in Euclid.

Because of this result it is clear that there must be enormously large numbers which are primes. But if a very large number is given, it is not at all an easy matter to decide if it is a prime. Unbelievably, this problem requires very sophisticated methods for its solution. Even in early times large primes were a source of interest. For instance, **Fermat** believed that all numbers of the form

$$A_n = 2^{2^n} + 1, \qquad n = 0, 1, 2, 3, \ldots$$

are primes. This is true for n upto 4 but false for $n = 5$ as Euler verified later. Nowadays testing for primality as well as factorization of large numbers is attracting a lot of attention because of the relationship of these questions to the problem of constructing and breaking codes.

NOTES AND EXERCISES

I have mentioned that Euclid's axioms are rather subtle in many ways. For instance, the plane in Euclid's geometry is not the surface of the earth but an ideal plane where lines stretch out to infinity in both directions. This is rather remarkable, because even though the origins of geometry go back ultimately to measurements carried out on the surface of the earth, nevertheless, when Euclid idealized them, everything took place in the ideal plane.

Among Euclid's axioms is the so-called *parallel postulate*, which stood out from the rest because it was very hard to accept it on any intuitive basis. This postulate amounts to saying that if a line ℓ is given and a point P not on it is also given, there is exactly one line through P which does not meet ℓ. Even Euclid must have been uncomfortable with it as may be seen from the fact that he made an effort to prove many propositions without the aid of the parallel postulate, and started using it only when it became absolutely necessary. His successors tried in vain to deduce this axiom from the other axioms, and this effort, lasting several centuries, ultimately led to the creation of *non euclidean geometry*, by **Bolyai** (1802–1860) and **Lobachevsky** (1793–1856).

After the publication of Euclid's *Elements*, the axiomatic approach became the preferred way to develop any mathematical topic. It was felt that the axiomatic method represented an ideal of perfection, and that the success of any development was measured against that of Euclid's geometry. As mentioned before, even Euclid's axioms had to be enlarged and modified so that the deductions made by him were permissible. The main figures in this line of thought were **Pasch** (1843–1930) and **Hilbert** (1862–1943). Nowadays the axiomatic method has penetrated all parts of mathematics and is the only approach used.

As a way to recall what was learned in school about geometry try to work out the proofs of exercises 1–3 below in Euclidean geometry noting carefully all the earlier propositions on which these proofs are based. Can you locate in which of these and where the parallel postulate is used?

1. The diagonals of a rectangle are equal.

2. Let AB be a line and let BC and AD be perpendicular to AB on the same side of AB such that $BC = AD$. Then

    (a) The angles DCB and BCA are equal.

    (b) The above angles are both right angles.

3. The sum of the angles of a triangle is two right angles.

4. This and the next exercise lead to a proof of the infiniteness of the sequence of primes. Prove first that if n is any positive integer, either it is a prime, or else it is divisible by a prime. (*Sketch of argument* : If n is not a prime, then we can write $n = n_1 n_2$ where n_1 and n_2 are both different from 1 and n. So $1 < n_1 < n$ and argue again the same way for n_1. This procedure will end in at most n steps, usually much less.)

5. The number of primes is infinite.

    (*Sketch of Euclid's argument* : Otherwise, let $p_1, p_2, \ldots p_n$ be *all* the primes. Consider the number

    $$p_1 p_2 \ldots p_n + 1$$

    By exercise 4, it is either a prime or is divisible by a prime. This means it is either one of the p_i's or is divisible by one of them. But this number leaves the remainder 1 when divided by any p_i.)

6. If n is not a prime, show that it has a prime factor $\leq \sqrt{n}$. (This is very useful in testing the primality of numbers of moderate size.)

7. Verify that the numbers $2^{2^n} + 1$ are, for $0 \leq n \leq 5$,

$$3, 5, 17, 257, 65537, 4294967297$$

Try to verify that all except the last one are primes and show that 641 divides the last one.

ARCHIMEDES (287 B. C.–212 B. C.)

Archimedes was one of the greatest mathematicians of all time, and certainly one of the most celebrated. He was a native of the Greek town of Syracuse situated on the island of Sicily. His genius was universal, allowing him to make fundamental discoveries in physics as well as mathematics. He was also a great inventor and there are many legends surrounding his achievements. He was supposed to have devised powerful catapults which rained heavy objects on invading armies. He was reported to have burned ships threatening his country by focusing the sun's rays on them. Everyone knows about the story of his discovery of the fundamental principle of hydrostatics that a body when immersed in a liquid displaces an amount equal in volume to it; he is supposed to have discovered it when he immersed himself in the bath tub, and was so excited that he ran out without any clothes, shouting "Eureka, eureka" (I have discovered!). Even today when someone discovers something important it is often the case that the person shouts "Eureka". For example, when Gauss discovered the proof that every positive integer is a sum of three *triangular* numbers (the definition of what is meant by a triangular number need not concern us here), he denoted it in a diary[1] he kept of his discoveries as follows:

EUREKA. NUMBER=$\Delta + \Delta + \Delta$

Archimedes was a great geometer as well as an arithmetician. In geometry he proved many propositions about the circle, sphere, and other geometrical figures. In arithmetic and its relation to geometry he is perhaps most famous for his measurement of the circumference of a circle. It was he who introduced the notation π (the first letter of the Greek word denoting circumference) to denote the ratio of the circumference of a circle to its diameter, and made the famous approximation

$$3\frac{10}{71} < \pi < \frac{22}{7} = 3\frac{1}{7}$$

Everyone who has studied mathematics in school knows the approximation 22/7 for π.

DIOPHANTUS (c. 250 A. D)

Diophantus was a Greek mathematician whose work on Algebra and Arithmetic called *Arithmetika* exerted an enormous influence on his successors, especially the French Mathematician **Pierre de Fermat** who is generally regarded as the father of modern number theory. Diophantus' work is supposed to have contained 13 books, but for a long time only 6 of these were available in Greek. Then in the mid 1970's Arabic translations of 4 more of the books were discovered.

Diophantus was interested in solving equations involving one or more unknown variables, and he appears to have been the first to ask that the solutions be rational fractions. He was perhaps aware that the solutions may not be unique although he does not seem to have asked for the enumeration of all the solutions. Among very simple examples of such problems are the following.

---

[1] For a commentary and translation of Gauss's diary, see the article by J. J. Gray, *A commentary on Gauss's mathematical diary, 1796–1814, with an English translation*, Expositiones Mathematicae, 2(1984), pp. 97–130.

> How many ways there are to divide a flock of 25 sheep into groups of 5's and 2's?
>
> Find two numbers such that the square of either added to the other gives a square.

Such problems are known as *diophantine problems* and there have been many famous ones in history, such as

> Which numbers can be written as a sum of two squares?
>
> Can the sum of two cubes be a cube, a sum of two fourth powers be a fourth power, and so on?

The second of these is the famous **Fermat's Last Theorem** which was posed by Fermat. Fermat worked with a copy of Diophantus's Arithmetika edited by **Bachet** and had commented that

> I have a truly marvellous proof that this cannot be so, but there is too little space in the margin for me to give it here.

Entire theories of mathematics were constructed in the effort to solve this question. Indeed, one may say that modern number theory was born in an attempt to devise tools for the solution of this problem. The conventional wisdom is that Fermat was in error and did not possess a proof in the general case, although he had a perfectly legitimate proof for the case $n = 4$, while Euler proved the result for $n = 3$ much later. The problem was unsolved till 1994 when **Andrew Wiles**, an English mathematician working in Princeton University, stunned the world of mathematics by finding a proof. His proof of course is not from first principles and uses very sophisticated ideas from modern number theory and algebraic geometry, and is published in *Annals of Mathematics*, Vol 141 (1995), pp. 443–572.

We shall see later that it was only after the sixteenth century or so that algebraic notation as we now know slowly came into being. However one can try to rewrite the problems mentioned above in terms of equations. Thus the problem of dividing the 25 sheep into groups of 5's and 2's becomes the following. Let us write x for the number of groups of 5 sheep and y for the number of groups of 2 sheep. Then we must have the equation

$$5x + 2y = 25 \qquad (x, y \text{ are integers } > 0)$$

This equation has many solutions which can be found out by trial and error. Thus we have

$$x = 1, y = 10, \qquad x = 3, y = 5$$

are solutions. Usually in algebra, if there are 2 unknown quantities to be solved for, there will be two equations; here there is only one equation, and we are required to find all solutions, but with the additional requirement that x and y be both positive integers. This is an example of a *diophantine equation*. We shall see in the next chapter that mathematicians from India in the period from the 6th to the 12th centuries made a deep study of the diophantine equations

$$x^2 - Ny^2 = \pm 1$$

where N is a positive integer which is not a square and $(x, y$ are to be positive integers.

Turning to the next set of equations, the question of which squares are the sum of two squares, the equation is

$$x^2 + y^2 = z^2 \qquad (x, y, z \text{ are integers } > 0)$$

Every one is familiar with the solutions

$$x = 3, y = 4, z = 5, \qquad x = 5, y = 12, z = 13$$

Clearly, if x and y are the lengths of the sides of a right triangle that contain the right angle, then z is the length of the hypotenuse. Because of this connection with the Pythagoras theorem, the sets of positive integers (x, y, z) satisfying the equation

$$x^2 + y^2 = z^2$$

are called *Pythagorean triplets*. We shall look into them more closely in the next chapter. The ancients knew ways to generate these triplets where x, y, z are huge numbers.

NOTES AND EXERCISES

Pierre de Fermat (1601–1665) is generally regarded as the founder of modern *Number theory*, a branch of mathematics generally dealing with whole numbers and rational numbers, and diophantine equations. He developed systematic methods for studying certain classes of diophantine equations and discovered many remarkable and deep properties of their solutions. The reader who wants to know more about the history of number theory and the men such as Fermat who created it, should read the wonderful book [W] of **André Weil** (1906–), one of the greatest mathematicians and number theorists of this century.

Fermat's problem is to show that for $n \geq 3$ there are no solutions in positive integers x, y, z of the equation

$$x^n + y^n = z^n \tag{$*$}$$

Fermat proved this for $n = 4$ by showing more generally that there are no solutions in positive integers x, y, z of the equation

$$x^4 + y^2 = z^4$$

As mentioned above, **Euler** (1707–1783) solved the Fermat problem $(*)$ for $n = 3$. It was in the course of his unsuccessful attempt to solve $(*)$ for general n that **Kummer** (1810–1893) laid the foundations of modern algebraic number theory.

1. Show that for the equation $5x + 2y = 25$ the solutions given are the only ones.

2. Check by actual calculation that there are no solutions to the equation

$$x^3 + y^3 = z^3 \qquad (x, y, z \text{ are positive integers })$$

with $x \leq 5, y \leq 6$.

3. The Babylonians knew pythagorean triplets of huge sizes. Check that the following are pythagorean triplets:

$$x = 120, y = 119, z = 169$$

$$x = 3456, y = 3367, z = 4825$$

2

PYTHAGORAS AND THE
PYTHAGOREAN TRIPLETS

Pythagoras lived in the sixth century B. C. and was not only a mathematician but also a philosopher. He had many followers who were called Pythagoreans. He is of course most famous for his theorem about the squares on the sides of a right triangle. But the Pythagoreans also discovered *irrational numbers*. In geometrical language, if AB and CD are two segments, they are *commensurable* if there is a unit EF such that both AB and CD are multiples of EF, so that $AB : CD = m : n$ where m and n are positive integers. The discovery of the Pythagoreans, which was a watershed event in the history of numbers, was that the side and the diagonal of a square are not commensurable. If the side has unit length 1, this means that the diagonal, which is $\sqrt{2}$ in length, is not commensurable with the side, i.e., $\sqrt{2}$ is not a rational number. We shall say more about this a little later. The classical argument is as follows. Suppose if possible that $\sqrt{2}$ is a fraction m/n where we may assume that m and n have no common factors. Then by squaring the relation

$$\sqrt{2} = \frac{m}{n}$$

we get

$$m^2 = 2n^2$$

This shows that m^2 is even, hence that m is even. Thus we can write $m = 2m'$ where m' is a positive integer. The relation $m^2 = 2n^2$ now becomes $4m'^2 = 2n^2$ or

$$n^2 = 2m'^2$$

from which we conclude as we did earlier that n is even. Thus both m and n are even, contradicting the assumption we made that they have no common factor.

We turn to a discussion of Pythagorean triplets. These are triplets of whole numbers (a, b, c) such that $a^2 + b^2 = c^2$. We may therefore regard them as solutions in positive integers of the diophantine equation

$$x^2 + y^2 = z^2$$

Of course if (a, b, c) is a solution, so is (b, a, c). Examples that everyone knows are $(3, 4, 5)$ and $(5, 12, 13)$. Ancient Hindu manuals known as *Śulva sūtras* in which details of altar constructions are given, have revealed evidence of knowledge of such triplets among those people. Also tablets dating back to the Hammurapi dynasty contain such triplets. The most famous of these is the one known as "Plimpton 322" [vW1] [W]. This tablet contains the following triplets.

$$(120, 119, 169), (4800, 4601, 6649), (72, 65, 97)$$

$$(600, 481, 769), (2700, 1771, 3229), (90, 56, 106)$$

It is reasonable to conclude from the large sizes of these numbers that whoever knew these also knew some general method of constructing such triplets. If one uses some elementary number theory it is not too difficult to obtain a general formula that will generate all such triplets.

The Plimpton 322 tablet which contains a list of Pythagorean triplets has the numbers written in the so-called *sexagesimal* system. This is a variant of the way we normally write numbers in powers of 10. Thus the number

$$1246879$$

really stands for

$$1 \times 10^6 + 2 \times 10^5 + 4 \times 10^4 + 6 \times 10^3 + 8 \times 10^2 + 7 \times 10 + 9$$

Other systems were used in ancient civilizations. The sexagesimal system is based on powers of 60. For instance

$$2, 0$$

means the number

$$2 \times 60 + 0$$

so that the triplet

$$120; 119; 169$$

in ordinary (i.e., in powers of 10) notation becomes in the sexagesimal notation

$$2, 0; 1, 59; 2, 49$$

(The reason for a comma between the successive digits is because in each place one can go from 0 to 59 and so it is necessary to separate the various places clearly.) The Mayans worked in the scale of 20–the *vigesimal* system. In our own age, electronic technology uses the *binary* scale, namely the scale of powers of 2. In modern number theory, not only 2 but every prime number is used; numbers expressed in the scale of the prime number p are examples of what are called p-adic numbers.

The Plimpton 322 tablet referred to contains the following triplets:

a	b	c
120	119	169
3456	3367	4825
4800	4601	6649
13500	12709	18541
72	65	97
360	319	481
2700	2291	3541
960	799	1249
600	481	769
6480	4961	8161
60	45	75
2400	1679	2929
240	161	289
2700	1771	3229
90	56	106

Actually the numbers in the tablet were in the sexagesimal system and the table looked like this:

a	b	c
2,0	1,59	2,49
57,36	56,7	1,20,25
1,20,0	1,16,41	1,50,49
3,45,0	3,31,49	5,9,1
1,12	1,5	1,37
6,0	5,19	8,1
45,0	38,11	59,1
16,0	13,19	20,49
10,0	8,1	12,49
1,48,0	1,22,41	2,16,1
1,0	45	1,15
40,0	27,59	48,49
4,0	2,41	4,49
45,0	29,31	53,49
1,30	56	1,46

The large sizes of these numbers are clearly an indication that the Babylonians knew how to generate such triplets in a systematic manner. There is also a more subtle indication. Notice that if (a, b, c) is a triple and we multiply all of them by the same number k so that we get (ka, kb, kc), then this is again a pythagorean triplet, because

$$k^2a^2 + k^2b^2 = k^2(a^2 + b^2) = k^2c^2$$

Thus once we obtain one triplet we can generate all their multiples. This shows that it is enough to obtain all triplets (a, b, c) with the property that the numbers a, b, c do not have a common factor. Such triplets are called *primitive*. In the above tablet, all except the ones in line 11 and line 15 are primitive. This is a very interesting indication that there was a fairly sophisticated understanding of pythagorean triplets going back to the babylonians.

Is there a general formula that can be used to generate all primitive triplets? There is one, and its derivation needs a little more effort. We shall not discuss it here in full detail but give only the final formula and an indication of how it comes about. Suppose that (a, b, c) is a pythagorean triplet which is primitive. Suppose first that both a and b are even. Then $c^2 = a^2 + b^2$ is even, which makes c even. So 2 divides all three of them, violating our assumption of primitivity. On the other hand, suppose that a and b are both odd. Then a^2 and b^2 are both odd, so that c^2 is even, showing that c must be even. But a^2 and b^2 must both leave the remainder 1 when divided by 4 so that c^2 must leave the remainder 2 when divided by 4, a contradiction, because as c is even, c^2 is divisible by 4! Thus, one of a and b is even, the other is odd. By switching a and b we may assume that a is even.

We now use the identity

$$(2st)^2 + (s^2 - t^2)^2 = (s^2 + t^2)^2$$

which was known to the ancients, to conclude that if we give s and t integer values, then

$$a = 2st, b = s^2 - t^2, c = s^2 + t^2 \qquad (*)$$

is a pythagorean triplet. For b to be positive, we must require that $s > t$. If s and t are chosen to have no common factor and to be of opposite parity, namely, that one of them is odd and the other is even, then one can show that $(*)$ is a primitive pythagorean triplet with a even. It is also true conversely that every primitive pythagorean triplet (a, b, c) with a even is of this form, for suitable integers s, t such that

$s > t > 0$, s and t have no common factors and are of opposite parity .

This discussion shows that the theory of diophantine equations is a very subtle one. That people of very ancient times possessed techniques for writing down these triplets is a remarkable fact. For those readers who are a little more versed in the elements of prime factorizations of numbers, I have indicated in the exercises below how the proofs of the statements made above can be developed.

NOTES AND EXERCISES

Several new concepts have been mentioned in the text, and it is a good idea to work through a few simple exercises to become familiar with them. We begin with the technique of expressing any number in several scales.

1. Write the number (given in the usual scale of 10) 13678 in the binary scale.

2. Check that the triplets in the Plimpton 322 tablet are all pythagorean and that all except those in lines 11 and 15 are primitive. Verify also that the second tablet consists of the same numbers as in the first but expressed in the sexagesimal scale.

3. Find, for each triplet in the Plimpton 322 tablet, the values of s and t such that the formula $(*)$ in the text which generates that triplet; for the ones in lines 11 and 15, you should first divide the triplet by a common factor to make it primitive and then find the values of s and t.

4. Deduce from the description of primitive pythagorean triplets that

$$a = 2s, b = s^2 - 1, c = s^2 + 1$$

with s even give all primitive triplets with $c - b = 2$.

5. If k is a positive integer show that $(2k + 1)^2 = 4k^2 + 4k + 1$ and deduce that the square of an odd integer is not only odd but leaves the remainder 1 when divided by 4. Using a similar argument, show that the square of an integer not divisible by 3 leaves the remainder 1 when divided by 3.

For proving that the description given above of all primitive triplets with a even is correct one needs to know that every integer is a product of primes (repetitions allowed) and that this can be done in a unique manner if we disregard the order in which the primes appear in the factorization. It follows from this that if a prime divides u^2 then it must divide u, and that if u, v are integers with no common factor and uv is a square, then each of u and v must be a square. Assuming these one can prove the following (This is a difficult exercise at this level. For a nice introduction to number theory see the books [HW] and [S]).

6. The formula $(*)$ defines a primitive pythagorean triplet with a even if and only if s and t have no common factors and are of opposite parity. (Hint : Assume the restrictions on s, t and suppose that a prime p divides a, b, c. Show that p must be odd and divides $2s^2, 2t^2$, hence s^2, t^2, hence s, t. For the converse, if s, t have the same parity, show that 2 divides a, b, c, and that if a prime p divides s, t, then p divides a, b, c also.)

7. Show that all primitive pythagorean triplets with a even can be obtained by $(*)$. (Hint : First show that if a and b have no common factor and $a^2 + b^2 = c^2$, then b and c must be both odd so that we can write $(a/2)^2 = ((c + b)/2)((c - b)/2)$ and argue that $(c + b)/2 = s^2$ and $(c - b)/2 = t^2$.)

8. Show that $\sqrt{3}$ is an irrational number. (Hint : Proceed as in the case of $\sqrt{2}$. The key is to use exercise 5 to show that if m^2 is divisible by 3, then so is m.)

It is clear that as we try to extend the argument given for the irrationality of $\sqrt{2}$ to other numbers such as $\sqrt{5}, \sqrt{7}$, and so on, the method becomes more and more laborious and so one wants an argument that will work quickly in all cases. The next two exercises are designed to do that. To do them one must remember the facts about prime factorization that were mentioned above.

9. Let m, n be two positive integers. Prove that if m and n have no common factor, then n and m^2 cannot have a common factor. Deduce that if $n > 1$ and m and n have no common factor, n cannot divide m^2.

10. Show that if the positive integer k is not the square of another positive integer, it cannot be the square of a rational number either, so that $\sqrt{k}$ is an irrational number. (Hint : If $k = m^2/n^2$ where m, n are positive integers without a common factor, deduce from $kn^2 = m^2$ that n divides m^2 and use the previous exercise to show that n must be 1.)

3

ĀRYABHAṬA–BRAHMAGUPTA–BHĀSKARA

Āryabhaṭa (476–c. 550), **Brahmagupta** (c. 598–665), and later on, **Bhāskara** (c. 1114–1185), were the most famous members of the Indian school of astronomy and mathematics that flourished from the 6th through the 12th centuries (see the detailed account of Indian mathematics of these epochs in the excellent source book of Datta and Singh [DS]). The Indians knew how to solve first and second degree diophantine equations, described algorithms for their solutions, and had many results about triangles with rational sides, right angled and otherwise. The pinnacle of their achievement was the discovery of the method for solving the diophantine equation

$$X^2 - NY^2 = \pm 1$$

where N is a given positive integer which is not a square and x and y are also positive integers. This equation is generally known as **Pell's equation** (due to a mistaken attribution of Euler). The method developed for its solution by the Indian school is known as the *cakravāla* in Sanskrit. It means the *cyclic method* (*cakra* means the wheel in Sanskrit). It is not known who discovered this method, but Bhāskara has expounded it and it appears now that it was also known to **Jayadeva** who lived in the 11th century. One of the remarkable aspects of this equation is that unlike any equation encountered hitherto it has *an infinite number of solutions*. **Brahmagupta** already knew this fact and was in fact familiar with what is nowadays understood as the multiplicative structure of the set of solutions, allowing him to construct an infinity of solutions as soon as a single solution is known. **Fermat** was however the first to insist on a proof that this equation always has an infinity of solutions and to have asked for a *method that generates all the solutions*. We shall discuss this below.

The *cakravāla* always leads to a non trivial solution. Even though Bhāskara appears to have asserted this as a fact, neither he nor anyone else of the Indian school who followed him appears to have written what we may accept as a *proof* for this statement. Still, the Indian mathematicians worked out numerical cases of great difficulty and intricacy. In fact, Weil who has given a beautiful account of this problem and the contributions of the Indian school to its solution has this to say in his book [W] (p. 81):

What would have been Fermat's astonishment, if some missionary, just back from India, had told him that his problem has been successfully tackled there by native mathematicians almost six centuries earlier!

The first definitive treatment of Pell's equation was given by the great French mathematician **Lagrange** (1736–1813).

Nevertheless one should not minimize the Indian achievement because of the absence of any indication that they were concerned about proving that their method always led to a solution or that it gave all the solutions. The smallest solutions for a given N can be very huge, and the discovery of the *cakravāla* and the systematic treatment of difficult numerical cases suggest that the Indian school must have been convinced that their method works in all cases. For instance, for the equation

$$X^2 - 61Y^2 = 1$$

the smallest solution is

$$X = 1766319049, Y = 226153980$$

(see Weil's book cited earlier, p. 97). Weil goes on to say (see p. 24) that

to have devoloped the *cakravāla* and to have applied it successfully to such difficult numerical cases as $N = 61$ or $N = 67$ had been no mean achievement.

The equation described above is a special case of a whole family of equations

$$X^2 - NY^2 = m \qquad\qquad (P_m)$$

where m is an integer $\neq 0$, positive or negative. Given a solution (x, y) to (P_m) and a solution (z, t) to (P_n), it is possible to construct from them in an explicit manner a solution to (P_{mn}). In particular, by repeatedly applying this procedure, starting with one solution to (P_1) (in positive integers), one can construct an infinite number of solutions to (P_1). This process of composition was known to Brahmagupta who called it the *bhāvanā*. We shall discuss this in greater detail below.

If we have a solution (x, y) of (P_m), then, writing

$$\frac{x}{y} - \sqrt{N} = \frac{1}{y}\frac{x^2 - Ny^2}{x + y\sqrt{N}} = \frac{1}{y}\frac{m}{x + y\sqrt{N}}$$

we see that x/y is a good rational approximation to $\sqrt{N}$ whenever x and y are large in comparison to m. Thus, the easily verified calculations

$$265^2 - 3 \times 153^2 = -2, \qquad 1351^2 - 3 \times 780^2 = 1$$

give the approximations

$$\frac{265}{113}, \qquad \frac{1351}{780}$$

to $\sqrt{3}$ which go back to Archimedes and his work on approximations to π. Actually it is true that

$$\frac{265}{153} < \sqrt{3} < \frac{1351}{780}$$

which formed the starting point to his work on π.

Eves [E] gives an interesting but brief account of some of the discoveries of the Indian school (see [E], pp. 172–187). Among these are the results of Brahmagupta on cyclic quadrilaterals, namely, quadrilaterals inscribed in a circle:

(1) The area A of a cyclic quadrilateral with consecutive sides a, b, c, d and semi-perimeter s, is given by

$$A = \sqrt{s(s-a)(s-b)(s-c)(s-d)}$$

(2) The diagonals m, n of such a quadrilateral are given by

$$m^2 = \frac{(ab+cd)(ac+bd)}{ad+bc}, \qquad n^2 = \frac{(ac+bd)(ad+bc)}{ab+cd}$$

(3) If a, b, c, A, B, C are positive integers such that $a^2 + b^2 = c^2$ and $A^2 + B^2 = C^2$ then the cyclic quadrilateral having sides aC, cB, bC, cA (called a *Brahmagupta trapezium*) has rational area and diagonals, and the diagonals are perpendicular to each other.

BRAHMAGUPTA'S WORK

The procedure mentioned in the text that allows one to construct additional solutions if we have some solution is based on *Brahmagupta's identity*:

$$(x^2 - Ny^2) \times (z^2 - Nt^2) = (xz \pm Nyt)^2 - N(xt \pm yz)^2$$

This is just a straightforward verification (see exercise 1 below). It follows at once from the above identity that if (p, q) is a solution to (P_m) and (r, s) is a solution to (P_n), then $(pr \pm Nqs, ps \pm qr)$ is a solution to (P_{mn}). Write

$$(p, q) * (r, s) = (pr + Nqs, ps + qr)$$

We think of this as the *composition* of (p, q) and (r, s). This process was called the *bhāvanā* by Brahmagupta, the word meaning "production" in Sanskrit. It is easy to verify that if p, q, r, s are all positive integers, then $(pr + Nqs, ps + qr)$ is different from the previous two (see exercise 2 below). Given a nontrivial solution (p, q) to (P_1), the *bhāvanā* allows one to obtain an *infinite* sequence of solutions $(p_k, q_k)(k = 1, 2, 3, \ldots)$ to (P_1) by

$$(p_0, q_0) = (p, q) \qquad (p_k, q_k) = (p, q) * (p_{k-1}, q_{k-1})(k = 1, 2, \ldots).$$

It is easy to see that these are all distinct. Similarly if we have in addition a solution (r, s) to $(P_{\pm m})$ and define (r_k, s_k) by

$$(r_0, s_0) = (r, s), \quad (r_k, s_k) = (p, q) * (r_{k-1}, s_{k-1})$$

then these are all solutions to $(P_{\pm m})$ which are distinct (see exercise 3 below).

NOTES AND EXERCISES

1. Verify Brahmagupta's identity.

2. Verify that if p, q, r, s are all positive integers and $(P, Q) = (p, q) * (r, s)$, then

    $$P > p, r, \qquad Q > q, s$$

    and hence show that (P, Q) is distinct from the original two.

3. Show that

    $$r_k > r_{k-1}, s_k > s_{k-1}$$

    and deduce that the (r_k, s_k) are all distinct if p, q, r, s are positive integers, using the same method as in the previous exercise.

4. Verify that $(2, 1))$ is a solution to $X^2 - 3Y^2 = 1$. Hence verify that (p_k, q_k) are solutions where

    $$(p_1, q_1) = (2, 1), \qquad (p_k, q_k) = (2p_{k-1} + 3q_{k-1}, p_{k-1} + 2q_{k-1})$$

    Show that the first six of these solutions are

    $$(2, 1), (7, 4), (26, 15), (97, 56), (362, 209), (1351, 780)$$

    Also starting with the solution $(1, 1)$ to $X^2 - 3Y^2 = -2$ verify that

    $$(1, 1), (5, 3)(19, 11), (71, 41), (265, 153)$$

    are also solutions to the same equation. Hence deduce the approximations to $\sqrt{3}$,

    $$\frac{265}{153} < \sqrt{3} < \frac{1351}{780}$$

    that were used by Archimedes.

5. Try to prove the theorems of Brahmagupta stated in the text on cyclic quadrilaterals (for more details, see [E], problem studies **7.9** and **7.10**).

CAKRAVĀLA

The *cakravāla* is an algorithm for solving $(P_{\pm 1})$ developed by the Indian school of mathematicians and astronomers between the 6th and 12th centuries ([W], [DS] (Part II)). We shall describe this algorithm a little later and illustrate it with some examples. While the algorithm is straightforward, one can often reduce the number of steps if one takes some short cuts. These short cuts, based on the *bhāvanā*, go back to Brahmagupta although he did not possess the algorithm. They show that if somehow one can come up with a solution in positive integers to (P_m) where $m = -1, \pm 2, \pm 4$, then one can get a solution in positive integers for (P_1), and hence an infinity of solutions by the *bhāvanā*. We shall give these rules now. For simplicity of notation let us write $(a, b; m)$ to mean that (a, b) is a solution to (P_m), i.e., $a^2 - Nb^2 = m$. In the boxes below p, q denote positive integers.

I. $(u, v; \pm 1) \implies (2u^2 \mp 1, 2uv; 1)$

II. $(p, q; \pm 2) \implies (p^2 \mp 1, pq; 1)$

III. $(p, q; 4), p$ even $\implies \left(\frac{1}{2}p^2 - 1, \frac{1}{2}pq : 1\right)$

$(p, q; 4), p$ odd $\implies \left(\frac{1}{2}p(p^2 - 3), \frac{1}{2}q(p^2 - 1); 1\right)$

IV. $(p, q; -4), p$ even $\implies \left(\frac{1}{2}(p^2 + 2), \frac{1}{2}pq; 1\right)$

$(p, q; -4), p$ odd $\implies (P, Q; 1)$

where

$$P = \frac{1}{2}(p^2 + 2)\{(p^2 + 1)(p^2 + 3) - 2\}, \qquad Q = \frac{1}{2}pq(p^2 + 1)(p^2 + 3)$$

For I we note that by forming the composition of (u, v) with itself we get $(U, V; 1)$ where

$$U = u^2 + Nv^2, \qquad V = 2uv$$

But
$$u^2 + Nv^2 = 2u^2 - (u^2 - Nv^2) = 2u^2 \mp 1$$

For II by composing (p, q) with itself we get $(P, Q; 4)$ where

$$P = p^2 + Nq^2 = 2p^2 - (p^2 - Nq^2) = 2(p^2 \mp 1), Q = 2pq$$

Now we use the principle that if

$$a^2 - Nb^2 = m^2$$

then

$$\left(\frac{a}{m}\right)^2 - N\left(\frac{b}{m}\right)^2 = 1$$

so that, taking $m = 2, a = P, b = Q$ we get

$$(p^2 \mp 1, pq; 1)$$

For III the same principle gives $(p/2, q/2; 1)$ and so, by I, we have, taking $u = p/2, v = q/2$,

$$\left(\frac{p^2}{2} - 1, \frac{pq}{2}; 1\right)$$

If p is even we stop here as the numbers above are positive integers. If p is odd, we compose this last with $(p/2, q/2)$ again to get $(P, Q; 1)$ where

$$P = \frac{p}{2}\left(\frac{p^2}{2} - 1\right) + N\frac{pq^2}{4} = \frac{p}{2}\left(\frac{p^2}{2} - 1\right) + p\frac{p^2 - 4}{4} = \frac{1}{2}p(p^2 - 3)$$

$$Q = \frac{q}{2}\left(\frac{p^2}{2} - 1\right) + \frac{p^2 q}{4} = \frac{q}{2}(p^2 - 1)$$

We now come to IV. As $(p, q; -4) \implies (p/2, q/2; -1)$ we compose this with itself to get, using I again,

$$\left(\frac{1}{2}(p^2 + 2), \frac{1}{2}pq; 1\right)$$

and we are done if p is even as the numbers above are positive integers. For p odd we take this last pair and compose it with itself twice. Now if we start with $(u, v; 1)$ then

$$(u, v) * (u, v) * (u, v) = (2u^2 - 1, 2uv) * (u, v)$$
$$= (2u^3 - u + 2Nuv^2, 4u^2 v - v)$$
$$= (4u^3 - 3u, 4u^2 v - v)$$

Taking $u = (1/2)(p^2 + 2), v = pq/2$ we get, after a straightforward calculation, the formula in IV.

Examples. 1. $N = 13.$ We have

$$(11)^2 - 13 \times 3^2 = 4$$

so that $(11, 3; 4)$. Rule III (when p is odd) of Brahmagupta now gives $(P, Q; 1)$ where

$$P = \frac{1}{2}11 \times 118 = 649, \qquad Q = \frac{1}{2}3 \times 120 = 180$$

so that

$$X = 649, \qquad Y = 180$$

is a positive integral solution of

$$X^2 - 13Y^2 = 1$$

2. $N = 61.$ A calculation shows that

$$(39)^2 - 61 \times 5^2 = -4$$

so that

$$(39, 5; -4)$$

Rule IV (with p odd) now gives, after a quick calculation using a hand calculator, $(P, Q; 1)$, where

$$P = 1523 \times \frac{1522 \times 1524 - 2}{2} = 1766319049$$

$$Q = \frac{1}{2}39 \times 5 \times 1522 \times 1524 = 226153980$$

Thus

$$X = 1766319049, \qquad Y = 226153980$$

is a positive integral solution of

$$X^2 - 61Y^2 = 1$$

DESCRIPTION OF THE CAKRAVĀLA We shall now describe the algorithm of *cakravāla*, in a form that is a variant of the original one as described in [W] (see p. 23). During this discussion which is somewhat more sophisticated than what we have seen hitherto, we shall use some properties of congruences. If m is a non zero integer and a, b are integers, we shall say that a *is congruent to b modulo m*, in symbols

$$a \equiv b(\bmod m),$$

if $a - b$ is divisible by m; the set of all such a for given b is a *congruence class*. Congruences can be manipulated like equalities. For the theory of congruences and their many applications, see the books [HW] and [S].

The objective is to find a non trivial solution of the equation

$$X^2 - NY^2 = 1$$

where N is a positive integer that is not a square. The idea is to start with some

$$(p_0, q_0; m_0) \qquad (m_0 = p_0^2 - N)$$

and construct in succession

$$(p_i, q_i; m_i) \qquad (i = 1, 2, \ldots)$$

The algorithm asserts that in a finite number of steps we will reach

$$(p_k, q_k; 1)$$

We start with p_0 such that p_0^2 *is as close to N as possible.* Then we have

$$(p_0, 1; m_0), \qquad m_0 = p_0^2 - N$$

If we now take some x_1 and form $(p_0, 1) * (x_1, 1)$, we get

$$(p_0 x_1 + N, p_0 + x_1; m_0(x_1^2 - N)) \tag{0}$$

The idea is now to choose x_1 so that $p_0 + x_1$ is divisible by $|m_0|$. This is of course always possible, and x_0 is not unique but is determined only upto an additive multiple of $|m_0|$. We exploit this ambiguity in x_1 to choose it so that

$$x_1^2 \text{ is as close to } N \text{ as possible} \tag{1_1}$$

(note that $m_0 < 0$). We now observe the key fact that $x_1^2 - N$ is divisible by m_0. For,

$$x_1^2 - N = x_1^2 - p_0^2 + m_0 = (p_0 + x_1)(x_1 - p_0) + m_0$$

is divisible by m_0 as $p_0 + x_1$ is so by choice of x_1. Moreover, as

$$p_0 x_1 + N = p_0(p_0 + x_1) - m_0$$

it follows that $p_0 x_1 + N$ is also divisible by m_0. Thus we get from (0),

$$(p_1, q_1; m_1)$$

where ($|x|$, the absolute value of x, is defined as x if $x \geq 0$ and $-x$ if $x < 0$)

$$p_1 = \frac{p_0 x_1 + N}{|m_0|}, \quad q_1 = \frac{p_0 + x_1}{|m_0|}, \quad m_1 = \frac{x_1^2 - N}{m_0} \tag{1_2}$$

Suppose now that we have formed $(p_i, q_i; m_i)$ where

$$p_i = \frac{p_{i-1} x_i + N q_{i-1}}{|m_{i-1}|}, \quad q_i = \frac{p_{i-1} + x_i q_{i-1}}{|m_{i-1}|}, \quad m_i = \frac{x_i^2 - N}{m_{i-1}} \tag{i}$$

and p_i, q_i, m_i are integers. We take the composition

$$(p'_{i+1}, q'_{i+1}) = (p_i, q_i) * (x_{i+1}, 1)$$

for suitable x_{i+1}. Now

$$p'_{i+1} = p_i x_{i+1} + N q_i, \quad q'_{i+1} = p_i + q_i x_{i+1}, \quad {p'_{i+1}}^2 - N {q'_{i+1}}^2 = m_i(x_{i+1}^2 - N)$$

We now want to choose x_{i+1} so that q'_{i+1} is divisible by m_i. But, and this is the observation of Weil mentioned above,

$$p_i - x_i q_i = q_{i-1} \frac{N - x_{i-1}^2}{|m_{i-1}|} = \pm m_i q_{i-1}$$

so that to make q'_{i+1} divisible by $|m_i|$ all we have to do is to choose x_{i+1} such that

$$x_{i+1} \equiv -x_i (\bmod |m_i|)$$

Thus one can always choose x_{i+1} so that $p_i + x_{i+1} q_i$ is divisible by $|m_i|$, and the choice is still possible within a congruence class modulo $|m_i|$. Once again we fix the choice of x_{i+1} so that

$$x_{i+1}^2 \text{ is as close to } N \text{ as possible} \qquad ((i+1)_1)$$

Once x_{i+1} is chosen, it turns out, as it did at the start of this discussion, that p'_{i+1} and $(x_{i+1}^2 - N)$ are divisible by m_i. Indeed, modulo $|m_i|$,

$$p'_{i+1} \equiv -p_i x_i + N q_i = \frac{p_{i-1}(N - x_i^2)}{m_{i-1}} = \pm p_{i-1} m_i \equiv 0$$

while

$$x_{i+1}^2 - N \equiv x_i^2 - N = \pm m_{i-1} m_i \equiv 0$$

Thus, setting

$$m_{i+1} = \frac{x_{i+1}^2 - N}{m_i} \qquad ((i+1)_2)$$

and

$$p_{i+1} = \frac{p_i x_{i+1} + N q_i}{|m_i|}, \quad q_{i+1} = \frac{p_i + q_i x_{i+1}}{|m_i|} \qquad ((i+1)_3)$$

we get

$$(p_{i+1}, q_{i+1}; m_{i+1})$$

These observations establish that the algorithm is defined without any ambiguity.

It remains to make precise what we mean by the requirement u^2 *be as close to N as possible*. One way is to always insist that

$$u < \sqrt{N} \qquad \text{and } N - u^2 \text{ is as small as possible}$$

This means that we choose p_0 so that

$$p_0 < \sqrt{N} < p_0 + 1$$

and the x_i so that

$$x_{i+1} \equiv -x_i \pmod{|m_i|} \qquad x_{i+1} < \sqrt{N} < x_{i+1} + |m_i| \qquad (i \geq 0)$$

Then one can show that

$$|m_i| < 2\sqrt{N}, \qquad |x_i| < \sqrt{N}$$

for all i. In fact, at the start, we have

$$x_1 < \sqrt{N} < x_1 + |m_0|$$

But, as $0 < p_0 < \sqrt{N}$, we have,

$$|m_0| = N - p_0^2 = (\sqrt{N} + p_0)(\sqrt{N} - p_0) < 2\sqrt{N}$$

and so

$$|x_1| < \sqrt{N}$$

Now

$$m_1 = \frac{x_1^2 - N}{m_0} = -\frac{(\sqrt{N} + x_1)(\sqrt{N} - x_1)}{m_0}$$

so that, using

$$|\sqrt{N} + x_1| < 2\sqrt{N}, \qquad |\sqrt{N} - x_1| < |m_0|$$

we get

$$|m_1| < \frac{2\sqrt{N}|m_0|}{|m_0|} = 2\sqrt{N}$$

Suppose we know that $|m_i| < 2\sqrt{N}$. We then have

$$x_{i+1} < \sqrt{N} < x_{i+1} + |m_i|$$

showing as before that

$$|x_{i+1}| < \sqrt{N}$$

and thence

$$m_{i+1} = \frac{x_{i+1}^2 - N}{m_i} = -\frac{(\sqrt{N} + x_{i+1})(\sqrt{N} - x_{i+1})}{m_i}$$

so that as before,

$$|m_{i+1}| < \frac{2\sqrt{N}|m_i|}{|m_i|} = 2\sqrt{N}$$

In other words, all the moduli m_i are within $(-2\sqrt{N}, +2\sqrt{N})$. Since the m_i are integers, it follows that they keep on repeating themselves in cycles. This is perhaps why the Indians called this process by the name *cakravāla* or the *cyclic method*.

The algorithm may be summarized as follows.

1. Find p_0, q_0, m_0 so that

$$p_0 < \sqrt{N} < p_0 + 1, \quad q_0 = 1, \quad m_0 = p_0^2 - N$$

2. Find x_1 such that

$$p_0 + x_1 \equiv 0 \ (\mathrm{mod} \ |m_0|), \quad x_1 < \sqrt{N} < x_1 + |m_0|$$

3. Define

$$p_1 = \frac{p_0 x_1 + N}{|m_0|}, \quad q_1 = \frac{p_0 + x_1}{|m_0|}, \quad m_1 = \frac{x_1^2 - N}{m_0}$$

4. If $p_i, q_i, m_i = p_i^2 - N q_i^2$ have been found, find x_{i+1} such that

$$p_i + x_{i+1} q_i \equiv 0 \ (\mathrm{mod} \ |m_i|), \quad x_{i+1} < \sqrt{N} < x_{i+1} + |m_i|$$

5. Define

$$p_{i+1} = \frac{p_i x_{i+1} + N q_{i+1}}{|m_i|}, \quad q_{i+1} = \frac{p_i + q_i x_{i+1}}{|m_i|}, \quad m_{i+1} = \frac{x_{i+1}^2 - N}{m_i}$$

6. Keep going till you reach

$$m_k = p_k^2 - N q_k^2 = 1$$

Of course there is no need to say that this is an astonishing algorithm. It is absolutely not obvious that this procedure ever terminates with some $m_k = 1$, or even that the p_i and q_i are all positive integers. If the algorithm leads to an $m_k = 1$, we shall have the solution

$$X = p_k, \qquad Y = q_k$$

for

$$X^2 - N Y^2 = 1$$

The *cakravāla* as the Indians used it, differed from the above in one respect; the choice of x_{i+1} in the congruence class of $-x_i$ modulo $|m_i|$ was required to be made in such a way as to minimize $N - x_{i+1}^2$ (see [DS], pp. 161–181, Part II). I have followed Weil [W] in describing the variant as well as the argument for the existence of x_{i+1}; for Bhāskara and his successors the argument for the choice of x_{i+1} was a little more involved. The Indians also used the short cuts mentioned earlier, based on Brahmagupta's rules, to reduce the computation (the technology of writing was not advanced in India and so there was a premium on short calculations as well as brief expositions!); however these short cuts are not strictly necessary because one can show that the cakravāla, applied in a straightforward manner, always leads to a solution, and in fact, to all the solutions. Finally, as I have already mentioned, there is no indication that the Indians had a proof that this algorithm will always lead to a solution. As we shall see presently, they worked out very difficult numerical cases like $N = 61, 67$ which suggest that they had great conviction that this process will always lead to a solution. It was not till 1768–1770, when Lagrange published his deep studies of the theory of *continued fractions* of quadratic irrationalities, in which he proved their periodicity and gave a definitive treatment of the equations $X^2 - NY^2 = \pm 1$, that it became possible to establish that the *cakravāla* always led to a solution of $X^2 - NY^2 = 1$, and further that it gave also the fundamental solution, namely, that one by composing which with itself one obtains *all* positive integral solutions. The treatment of these things is beyond the scope of this little volume. However in the exercises we shall indicate the proof that the p_i and q_i are strictly increasing sequences of positive integers, and that they are *precisely* the convergents to the continued fraction expansion of $\sqrt{N}$, and hence that they contain all the solutions of the Pell's equation.

Example. $N = 13$. We have $p_0 = 3$ as 9 is the largest square less than 13. Then $m_0 = -4$ and we have $(3, 1; -4)$. Already by rule IV we can get the solution. But we shall not take this short cut but proceed strictly as required by the *cakravāla*. Then $0 \le x_1 \le 3$ and $3 + x_1$ is to be divisible by 4 giving $x_1 = 1$. So

$$p_1 = 4, \quad q_1 = 1, m_1 = 3$$

giving us $(4, 1; 3)$. Then $1 \le x_2 \le 3$ and $4 + x_2$ is to be divisible by 3, so that $x_2 = 2$. So

$$p_2 = 7, \quad q_2 = 2, \quad m_2 = -3$$

giving us $(7, 2; -3)$. Proceeding in this manner we get in succession

$$(11, 3; 4), \qquad (18, 5; -1), \qquad (119, 33; 4), \qquad (137, 38; -3)$$

$$(256, 71; 3), \qquad (393, 109; -4), \qquad (649, 180; 1)$$

We have already seen that

$$(649, 180)$$

is a solution in this case. We can also do this in the manner of Bhāskara, with the x_i chosen differently as explained.

NOTES AND EXERCISES

1. Treat the case $N = 61$ by the *cakravāla* without using short cuts, proceeding as in the example above, and show that one gets the solution

$$X = 1766319049, \qquad Y = 226153980$$

in either method.

2. Use the *cakravāla* to find likewise the following solutions:

   (1) $N = 67, X = 48842, Y = 5967$

   (2) $N = 103, X = 227528, Y = 22419$

   (3) $N = 97, X = 62809633, Y = 6377352$

The aim of the following exercises is to relate the cakravāla to the theory of simple continued fractions and develop a proof that the method leads to all solutions of Pell's equation.

3. Show that $m_i m_{i+1} = x_{i+1}^2 - N < 0$ and hence that the m_i are alternately positive and negative, starting with $m_0 < 0$.

4. Deduce from the formulae for p_{i+1}, q_{i+1} that

$$p_{i+1} q_i - q_{i+1} p_i = (-1)^i$$

   (Hint : Observe that $m_i = (-1)^{i-1}|m_i|$.)

5. Show that $p_1 = 1 + p_0 q_1$ and further that $-p_0 < \sqrt{N} - |m_0|$. Hence deduce that $q_1 > 0$ and $p_1 > p_0 > 0$. (Hint : The formula for p_1 is a straightforward calculation. Next note that $(\sqrt{N} + p_0)/|m_0| = 1/(\sqrt{N} - p_0) > 1$, proving the second assertion. Now x_1 is to be chosen in the form $-p_0 + c|m_0|$ for some integer c and satisfying $\sqrt{N} - |m_0| < x_1 < \sqrt{N}$, and then $c = q_1$. Show from the above inequality that $c > 0$.)

6. Show from the formulae $((i + 1))$ that for all $i \geq 0$,

$$p_{i+1} = a_{i+1} p_i + p_{i-1}, \qquad q_{i+1} = a_{i+1} q_i + q_{i-1}$$

   where

$$a_{i+1} = \frac{x_{i+1} + x_i}{|m_i|}$$

   with the conventions $q_0 = 1, p_{-1} = 1, q_{-1} = 0, a_0 = p_0, a_1 = q_1$. Hence deduce that

$$p_{i+1} q_{i-1} - q_{i+1} p_{i-1} = (-1)^{i-1} a_i$$

7. Show that $-x_i < \sqrt{N} - |m_i|$ and hence deduce that a_i is a positive integer for all i. Deduce from exercise 5 that the p_i and q_i are strictly increasing sequences of positive integers. (Hint: We have $x_{i+1} = -x_i + c|m_i|$ for an integer c and then $a_{i+1} = c = (x_{i+1} + x_i)/|m_i|$. Assuming the inequality, argue that c has to be > 0 because x_{i+1} has to be $> \sqrt{N} - |m_i|$. For proving the inequality, observe that $(\sqrt{N} + x_i)/|m_i| = |m_{i-1}|/(\sqrt{N} - x_i) > 1$ because x_i satisfies the inequality $0 < \sqrt{N} - x_i < |m_{i-1}|$.)

8. Fix $i \geq 0$ and let

$$\xi = \frac{\sqrt{N} q_i - p_i}{p_{i-1} - \sqrt{N} q_{i-1}}$$

   Show that

$$a_{i+1} = \left[\frac{1}{\xi}\right]$$

   where $[u]$ is the largest integer $\leq u$. (Hint : Using the expressions for p_i and q_i in terms of p_{i-1}, q_{i-1}, show that $1/\xi = |m_{i-1}|/(\sqrt{N} - x_i) = (\sqrt{N} + x_i)/|m_i|$. So, if $t = [1/\xi]$, then t is determined by the inequalities $t < (\sqrt{N} + x_i)/|m_i| < t + 1$ which are equivalent to

$$-x_i + t|m_i| < \sqrt{N} < -x_i + t|m_i| + |m_i|$$

so that $x_{i+1} = -x_i + t|m_i|$. This shows that $t = a_{i+1}$.)

It follows from this exercise that *the a_i are precisely the denominators in the simple continued fraction expansion of $\sqrt{N}$*. More precisely, using the notation of [HW],

$$\sqrt{N} = [a_0, a_1, a_2, \ldots]$$

To see this, let $\sqrt{N} = [b_0, b_1, b_2, \ldots]$. Since $b_0 = [\sqrt{N}]$ it is clear that $b_0 = a_0$. Further, $b_1 = [1/(\sqrt{N}-p_0)] = [(\sqrt{N}+p_0)/|m_0|)]$ so that it is the integer t defined by $-p_0+t|m_0| < \sqrt{N} < -p_0 + t|m_0| + |m_0|$, showing that $t = a_1 = q_1$. Thus $b_1 = a_1$. Suppose now that $b_i = a_i, i \leq n$. Then

$$\sqrt{N} = [a_0, a_1, \ldots, a_{n-1}, a_n + \eta] = \frac{p_n + \eta p_{n-1}}{q_n + \eta q_{n-1}}$$

where

$$\frac{p_n}{q_n} = [a_0, a_1, \ldots, a_n]$$

Then

$$\eta = \frac{\sqrt{N}q_n - p_n}{p_{n-1} - \sqrt{N}q_{n-1}}$$

and so

$$b_{n+1} = \left[\frac{1}{\eta}\right] = a_{n+1}$$

from the last exercise.

The proofs of the basic results pertaining to the *cakravāla* will now be proved if we establish that *the convergents to the continued fraction of $\sqrt{N}$ contain all the positive integral solutions of the equation $X^2 - NY^2 = \pm 1$*. Now one can show that

$$x^2 - Ny^2 = \pm 1 \implies |(x/y) - \sqrt{N}| < \frac{1}{2y^2} \tag{*}$$

Indeed, if we take the plus sign,

$$\frac{x^2}{y^2} = N + \frac{1}{y^2} \implies |(x/y) - \sqrt{N}| = (x/y) - \sqrt{N} = \frac{1}{y(x + y\sqrt{N})} < \frac{1}{2y^2}$$

because $x + y\sqrt{N} > 2y$ (since $x/y > \sqrt{N} > 1$). Suppose that $x^2 - Ny^2 = -1$. If $x \geq y$ we have as before

$$|(x/y) - \sqrt{N}| = \sqrt{N} - (x/y) = \frac{1}{y(x + y\sqrt{N})} < \frac{1}{2y^2}$$

since $x + y\sqrt{N} \geq y + Y\sqrt{N} > 2y$. But x cannot be $< y$; for, if this were so,

$$-1 < y^2 - Ny^2 \implies Ny^2 < y^2 + 1 \implies N < 1 + \frac{1}{y^2} \implies N < 2$$

giving $N = 1$ which is absurd. It is now a basic result in the theory of simple continued fractions (see [HW], Theorem 184) that the inequality $(*)$ implies that x/y is a convergent to the simple continued fraction for $\sqrt{N}$, so that $x = p_k, y = q_k$ for some k. See also [W], [S].

A note on the determination of Brahmagupta's epoch. Because of the paucity and lack of reliability of historical records in ancient India it is of interest to enquire how the epoch when Brahmagupta lived was determined. Brahmagupta wrote a well known text on astronomy in which he had included his observations on the distant stars. In particular his data allow the determination of the right ascension of the star *Spica Virginis*, known in India as *Chitra*, to be $182° 45'$. Its actual right ascension in 1800 A. D. was $198° 40' 2''$. The difference, $15° 55' 2''$ is due to the annual precession of the star which is taken to be $47''$, corresponding roughly to the period of revolution of 26,000 years for the orbit of earth around the distant star. This difference thus gives a lapse of 1216 years, thus indicating that the book was written around 584 A. D. For more details see the account given in the translation by H. T. Colebrooke of the works of Brahmagupta and Bhāskara : *Algebra, with arithmetic and mensuration, from the Sanscrit of Brahmagupta and Bhāscara, London, 1817.*

4

IRRATIONAL NUMBERS : CONSTRUCTION

AND APPROXIMATION

The Greeks already knew that not all segments that arise in geometrical constructions have lengths which are rational fractions. Thus $\sqrt{2}$ is not rational, although it is the length of the diagonal of a square of side 1.

A very general proposition in Euclid shows how to construct a square whose area is equal to that of a given rectangle. If the sides of the rectangle are a and b, then the side of the square constructed will be $\sqrt{ab}$. If we take a to be 1, we see that if one has constructed a segment of length b then one has a construction for a segment of length $\sqrt{b}$. In this way one has the possibility of constructing segments of lengths such as

$$\sqrt{1+\sqrt{2}}\,,\sqrt{1+\sqrt{2+\sqrt{3}}}\,,\ldots$$

Slowly mathematicians began to realize that these numbers which are not rational fractions, the so-called *irrational* numbers, are very important and it is necessary to find rational (for example, decimal) approximations to them.

The construction of $\sqrt{ab}$ by geometrical means can be carried out in many ways. One of the simplest is to assume $a > b$ (as we may) and use the identity

$$ab = \left(\frac{a+b}{2}\right)^2 - \left(\frac{a-b}{2}\right)^2$$

This shows that if we construct a right triangle ABC with hypotenuse AB equal to $(a+b)/2$ and one side AC equal to $(a-b)/2$, then the other side is $\sqrt{ab}$. To construct the triangle we proceed as follows. We draw a line AB of length $(a+b)/2$ and a semicircle above it with AB as a diameter. With A as center we draw a circular arc of radius $(a-b)/2$ and let C be the point where this arc meets the semicircle. Then the angle $< ACB$ is a right angle and AC has length $(a-b)/2$, so that BC has length $\sqrt{ab}$.

We have seen that in the work of Archimedes there appeared approximations to irrational numbers. Clearly it was understood intuitively that these numbers which came out of geometrical constructions were every bit as real as the rational numbers and that it was necessary to find good rational approximations for them. The classical and well-known algorithm for finding the square root of an integer is a systematic process for finding decimal approximations of arbitrary closeness to the square root concerned. Approximations for the cube roots were also developed.

NOTES AND EXERCISES

1. Prove the following method for constructing $\sqrt{ab}$. Assume $a > b$ and take a segment AC of length a and a point B in it such that $BC = b$. Extend AC to D so that $BD = a$ again. Using compass construct an isosceles triangle with base AD and sides $AE = DE = a$. Show that the triangles EBC and DEB are similar. Hence deduce that if x is the length BE, then

$$\frac{x}{b} = \frac{a}{x}$$

which gives

$$x = \sqrt{ab}.$$

Even in very early times it was noted that one can manipulate irrational numbers in the same manner as rational numbers. The next exercise is an example of this, attributed to Bhāskara in [E] (cf. p. 184).

2. Prove the two identities

$$\sqrt{a \pm \sqrt{b}} = \sqrt{(a + \sqrt{a^2 - b})/2} \pm \sqrt{(a - \sqrt{a^2 - b})/2}$$

3. The basic approximation in computing square roots is $\sqrt{a^2 + b} \sim a + b/2a$. Prove that

$$a \pm \frac{b}{2a} > a^2 \pm b$$

Use this repeatedly to get

$$\sqrt{2} \sim 1 + \frac{1}{3} + \frac{1}{12} - \frac{1}{408}$$

(Hint: The first application gives $3/2 > \sqrt{2}$. Then $2 = (3/2)^2 - (1/4)$ and so a second application gives $3/2 - 1/12 > \sqrt{2}$. A third application gives the required result.)

APPROXIMATIONS TO π

Throughout the history of mathematics, going back to the most ancient times, the number π, which denotes the ratio of the circumference of a circle to its diameter, has exerted a tremendous fascination and generated surprisingly profound and fertile questions.

Among the oldest problems involving π is that of **squaring a circle**, namely, to construct by geometrical means (which means using straightedge and compass only) a square whose area is equal to that of a given circle. Belonging to the same genre is the problem of constructing by geometrical means a line segment whose length is equal to the circumference of a given circle. The history of π resonates with some of the greatest names in mathematics. Here is a sample.

$$3\tfrac{10}{71} < \pi < 3\tfrac{1}{7} \text{ (ARCHIMEDES)}$$

$$\pi \approx \tfrac{62832}{20000} = 3.1416 \text{ (ĀRYABHAṬA)}$$

$$\tfrac{\pi}{4} = 1 - \tfrac{1}{3} + \tfrac{1}{5} - \tfrac{1}{7} \ldots \text{ (LEIBNIZ)}$$

$$\tfrac{\pi^2}{8} = 1 + \tfrac{1}{3^2} + \tfrac{1}{5^2} + \tfrac{1}{7^2} \ldots \text{ (EULER)}$$

A systematic way to construct approximations to π is to observe that if p_N^+ and p_N^- are the perimeters of the regular polygons P_N^+ and P_N^- respectively with N sides which respectively circumscribe and are inscribed in the given circle of radius 1, then

$$\tfrac{1}{2}p_N^- < \pi < \tfrac{1}{2}p_N^+$$

This is the method Archimedes used in his famous *Measurement of a circle*. For certain special values of N the $p_N^\pm$ can be estimated. For instance, Archimedes used $N = 96$ while Āryabhaṭa seems to have used $N = 384$. However a straightforward attempt to calculate these numbers will run into problems. We shall discuss briefly the famous derivation of Archimedes of his result

$$\frac{22}{7} > \pi > 3\frac{10}{71}$$

(See [H—A], *Measurement of a circle*, pp. 91–98).

THE APPROXIMATIONS OF ARCHIMEDES TO π

The idea is to start with a regular hexagon, inscribed or circumscribed, and successively estimate the sides of the regular polygons with $12, 24, 48, 96, \ldots$ sides. If we are willing to use a little trigonometry we can see clearly the method of Archimedes. For any regular polygon with N sides let θ_N be half of the angle subtended by the side at the center of the polygon. Then a look at the figure consisting of P_N^+, P_N^- and the circle will show that we have

$$\frac{1}{2}p_N^- = N \sin\theta_N, \qquad \frac{1}{2}p_N^+ = N \tan\theta_N$$

so that

$$N \sin \theta_N < \pi < N \tan \theta_N$$

The next step is to compute, or at least estimate, how $\sin \theta_N$ and $\tan \theta_N$ change when we go from N sides to $2N$ sides. This corresponds to changing θ_N to $(1/2)\theta_N$. The trigonometric identities we need are now

$$\frac{1}{\tan \theta/2} = \frac{1}{\sin \theta} + \frac{1}{\tan \theta}, \qquad \frac{1}{\sin \theta/2} = \sqrt{1 + \frac{1}{\tan^2 \theta/2}}$$

Archimedes works with equivalent results obtained by using Euclid's geometrical results. The appearance of the square root in the second of these formulae makes it clear that to get rational approximations we must replace the square roots by rational fractions close to their value. To use this technique systematically and in an iterative fashion we therefore need to have inequalities in place of the above equalities. The change from θ to $\theta/2$ is then given by the following: If we start with

$$\alpha < \frac{1}{\tan \theta} < a, \qquad \beta < \frac{1}{\sin \theta} < b$$

we get

$$\alpha + \beta < \frac{1}{\tan \theta/2} < a + b, \qquad \sqrt{1 + (\alpha + \beta)^2} < \frac{1}{\sin \theta/2} < \sqrt{1 + (a + b)^2}$$

Since we start with the hexagons the starting value of $\theta = \theta_6$ is $30°$. Then

$$\frac{1}{\tan \theta_6} = \sqrt{3}, \qquad \frac{1}{\sin \theta_6} = 2$$

Here Archimedes uses the approximations we have already discussed for $\sqrt{3}$, namely

$$\frac{265}{153} < \sqrt{3} < \frac{1351}{780}$$

So we have

$$\frac{571}{153} < \frac{1}{\tan \theta_{12}} < \frac{2911}{780}$$

The estimates for $\sin \theta_{12}$ follow from this but we have to use the estimates

$$\sqrt{349450} > 591\frac{1}{8}, \qquad \sqrt{9082321} < 3013\frac{3}{4}$$

to get

$$\frac{591\frac{1}{8}}{153} < \frac{1}{\sin\theta_{12}} < \frac{3013\frac{3}{4}}{780}$$

Thus

$$\frac{571}{153} < \frac{1}{\tan\theta_{12}} < \frac{2911}{780}, \qquad \frac{591\frac{1}{8}}{153} < \frac{1}{\sin\theta_{12}} < \frac{3013\frac{3}{4}}{780}$$

The next step is an iteration of this. For the tangent it is immediate :

$$\frac{1162\frac{1}{8}}{153} < \frac{1}{\tan\theta_{24}} < \frac{5924\frac{3}{4}}{780} = \frac{1823}{240}$$

The estimates for the sine involve using the inequalities

$$\sqrt{1373943\frac{33}{64}} > 1172\frac{1}{8}, \qquad \sqrt{3380929} < 1838\frac{9}{11}$$

and we obtain

$$\frac{1172\frac{1}{8}}{153} < \frac{1}{\sin\theta_{24}} < \frac{1838\frac{9}{11}}{240}$$

So

$$\frac{1162\frac{1}{8}}{153} < \frac{1}{\tan\theta_{24}} < \frac{1823}{240}, \qquad \frac{1172\frac{1}{8}}{153} < \frac{1}{\sin\theta_{24}} < \frac{1838\frac{9}{11}}{240}$$

Proceeding in the same way we get

$$\frac{2334\frac{1}{4}}{153} < \frac{1}{\tan\theta_{48}} < \frac{3661\frac{9}{11}}{240} = \frac{1007}{66}$$

and using the inequalities

$$\sqrt{5472132\frac{1}{16}} > 2339\frac{1}{4}, \qquad \sqrt{1018405} < 1009\frac{1}{6}$$

we get

$$\frac{2339\frac{1}{4}}{153} < \frac{1}{\sin\theta_{48}} < \frac{1009\frac{1}{6}}{66}$$

So

$$\frac{2334\frac{1}{4}}{153} < \frac{1}{\tan\theta_{48}} < \frac{1007}{66}, \qquad \frac{2339\frac{1}{4}}{153} < \frac{1}{\sin\theta_{48}} < \frac{1009\frac{1}{6}}{66}$$

For the last step we have

$$\frac{4673\frac{1}{2}}{153} < \frac{1}{\tan\theta_{96}} < \frac{2016\frac{1}{6}}{66}, \qquad \frac{1}{\sin\theta_{96}} < \frac{2017\frac{1}{4}}{66}$$

In deriving the second of these we have used

$$\sqrt{4069284\frac{1}{36}} < 2017\frac{1}{4}$$

Finally

$$96\tan\theta_{96} < \frac{96 \times 153}{4673\frac{1}{2}} < 3\frac{1}{7}$$

and

$$96\sin\theta_{96} > \frac{96 \times 66}{2017\frac{1}{4}} > 3\frac{10}{71}$$

Thus

$$3\frac{10}{71} < \frac{1}{2}p_{96}^{+} < \pi < \frac{1}{2}p_{96}^{-} < 3\frac{1}{7}$$

The reader who checks through this chain of estimates will realize how delicate were the calculations of Archimedes. To add mystery to this, the starting estimates for $\sqrt{3}$ were written down by him without any comment!

Starting from the seventeenth century it became an interesting problem to compute π accurate to as many decimal places as possible. We have already mentioned Leibniz's formula for $\pi/4$ which is obtained by putting $x = 1$ in the famous series discovered by the Scottish mathematician **James Gregory** (1638–1675),

$$\arctan x = x - \frac{x^3}{3} + \frac{x^5}{5} - \dots$$

For $x = 1$ this series converges too slowly to be of use in calculating π accurately. One can take other values, or use identities to transform this series into a rapidly converging one, as we shall indicate in the exercises below. A very nice account of the history of some of these calculations is given in [E], pp. 96–102. The Indian mathematical genius **Ramanujan (1887–1920)** (for an interesting account of his life see *The man who knew infinity* by *Robert Kanigel*, Scribner's, 1991) used deep arithmetical and function theories to produce very rapidly converging series for π which have allowed it to be calculated upto millions of decimal places. Here is an example.

$$\frac{1}{2\pi\sqrt{2}} = \frac{1103}{99^2} + \frac{27493}{99^6}\frac{1}{2}\frac{1.3}{4^2} + \frac{53883}{99^{10}}\frac{1.3}{2.4}\frac{1.3.5.7}{4^2.8^2} + \cdots$$

Taking only the first term we get

$$\frac{1103}{99^2} = .11253953678\ldots$$

$$\frac{1}{2\pi\sqrt{2}} = .11253953951\ldots$$

Ramanujan gave many other approximations to π. Here are two which ought to be compared with the true value. The first one is an improvement of the approximation $\frac{355}{113}$ which is quite old.

$$\pi = 3.14159265358979323846264345\ldots$$

$$\pi \approx \frac{355}{113}\left(1 - \frac{0.0003}{3533}\right) = 3.14159265358979432\ldots$$

$$\pi \approx \left(97\tfrac{1}{2} - \tfrac{1}{11}\right)^{1/4} = 3.14159265265262\ldots$$

From classical times the problem of squaring the circle occupied mathematicians. One can show that any positive solution to the problem of squaring the circle will imply that π satisfies an equation of the form

$$X^n + A_1 X^{n-1} + \ldots A_n = 0$$

where $A_1, A_2, \ldots, A_n$ are rational numbers. Remarkably, **Carl Lindemann (1852—1939)** proved that π cannot satisfy any such algebraic equation. Numbers which satisfy polynomial equations with rational coefficients are called *algebraic numbers*; all other numbers are known as *transcendental numbers*. Thus Lindemann's theorem amounts to the statement that π is a transcendental number. The terminology suggests that these numbers are beyond geometric and other constructions of the most general kind. Surprisingly, the mathematician **Georg Cantor**

(1845—1918) showed that most numbers are transcendental. Nevertheless, to decide whether a number that arises explicitly, such as π, e (the base of the natural logarithms), or $2^{\sqrt{2}}$, is irrational or transcendental is always a very difficult question. That e, the base of the natural logarithms, is transcendental, was first proved by **Hermite** (1822–1901). Such questions have had a deep interest for mathematicians always, right down to the modern era. Hilbert asked in his famous address to the International Congress of mathematicians in Paris in 1900 whether a^b is transcendental if a and b are algebraic numbers, b is irrational, and $a \neq 0, 1$. This was settled in the affirmative by **A. O. Gel'fond**, and by **Th. Schneider**, in 1934. Such questions severely tax the resources of modern mathematics. For differences between algebraic and transcendental numbers in the closeness to which they can be approximated by rational numbers, see [HW] [S].

NOTES AND EXERCISES

1. Verify the inequalities used in the Archimedes' approximation.

2. Prove the trigonometric identities used in the Archimedes' approximation.

To obtain an insight into the approximations for the square roots in Archimedes's work, see [H–A] (Introduction, pp. lxxxiv–xc). The basic inequalities are

$$a \pm (b/2a) > \sqrt{a^2 \pm b} > a \pm (b/2a \pm 1)$$

where $a, b > 0$, $a^2 + b^2$ is not a square, a^2 is the largest square less than $a^2 + b$, and in the inequality with the minus signs, $b < 2a - 1$ also. The discussion cited above makes the inequalities used very transparent.

3. Since $\tan(\pi/6) = 1/\sqrt{3}$, one can use Gregory's series to get

$$\frac{\pi}{6} = \sqrt{\frac{1}{3}} \left\{ 1 - \frac{1}{(3)3} + \frac{1}{(3^2)5} - \frac{1}{(3^3)7} \cdots \right\}$$

Calculate π to as many decimal places as you can, using this formula.

4. Writing $a(u) = \arctan u$, and using the formula

$$\tan(A + B) = \frac{\tan A + \tan B}{1 - \tan A \tan B}$$

deduce that

$$a(u) + a(v) = a\left(\frac{u + v}{1 - uv}\right)$$

$$a(u) + a(v) + a(w) = a\left(\frac{u + v + w - uvw}{1 - uv - vw - wu}\right)$$

5. Use the previous exercise to verify (see [E], pp. 96–102)

   (i) $\arctan \frac{1}{2} + \arctan \frac{1}{3} = \frac{\pi}{4}$

   (ii) $\arctan \frac{1}{2} + \arctan \frac{1}{5} + \arctan \frac{1}{8} = \frac{\pi}{4}$

   (iii) $4 \arctan \frac{1}{5} - \arctan \frac{1}{239} = \frac{\pi}{4}$

   (iv) $4 \arctan \frac{1}{5} - \arctan \frac{1}{70} + \arctan \frac{1}{99} = \frac{\pi}{4}$

   (Hint : For (iii) and (iv), show first that $3 \arctan \frac{1}{5} = \arctan \frac{37}{55}$)

6. Use the formulae in exercise 5 to calculate π accurately to several decimal places.

7. Show that

$$x = \sqrt[3]{3 + \sqrt{5 + \sqrt{2}}}$$

   is algebraic by constructing an equation of the 12th degree with integer coefficients satisfied by x.

Hilbert's famous address listed 23 problems and the problem of transcendentality of a^b is the 7th. Of these problems, Hermann Weyl had this to say:

   A mathematician who had solved one of them thereby passed on to the honors class of the mathematical community.

5

ARABIC MATHEMATICS

AL–KHWARIZMI : QUADRATIC EQUATIONS

Al–Khwarizmi (c. 780–c. 850) was an Arabic mathematician who lived in the 8th and 9th centuries. He was one of the most influential figures of his time and was a central figure in the separation of Algebra from Geometry and its subsequent development as an independent discipline. It is thought that the modern word *Algorithm* derives from his name. He classified quadratic equations and gave the method of solution for each of the species. Here is an example that typifies some of the problems he solved in his work.

One square and ten roots of the same amount to thirty-nine dirhems; that is to say, what must be the square, which, when increased by ten of its own roots, amounts to thirty-nine?

In modern notation this is the problem

Solve for X from the equation $X^2 + 10X = 39$

One of the greatest obstacles to the growth of Algebra was the absence of notation. To illustrate this let us take the equation that was studied by Al–Khwarizmi

$$X^2 + 10X = 39$$

A high school student today would solve it in the following steps:

$$X^2 + 10X + 25 = 39 + 25 = 64$$

$$(X + 5)^2 = 64 = 8^2$$

$$X + 5 = 8$$

$$X = 3$$

if we take only the positive solution. Let us compare this with Al–Khwarizmi's original solution. In studying it we should remember that Al–Khwarizmi is aware that he is solving not only this equation but all similar ones, and so the solution is written in a way that communicates this; he cannot do any better as he has no general notation.

> Halve the number of roots, which in this case yields five. This you multiply by itself; the product is twenty-five. Add this to thirty-nine; the sum is sixty-four. Now take the root of this, which is eight, and subtract from it half the number of roots, which is five. The remainder is three. This is the root you asked for.

This set of instructions applies to the equation

$$X^2 + 2BX = C$$

giving the sequence of steps

$$X^2 + 2BX + B^2 = C + B^2$$
$$(X + B)^2 = C + B^2$$
$$X + B = \sqrt{C + B^2}$$
$$X = \sqrt{C + B^2} - B$$

There is no need to dwell too much on how cumbersome were the processes available to people in those days in discovering and communicating mathematical ideas and calculations. It is a mystery that the Greeks faced and solved this problem in Geometry but not in Algebra. It was thus a great liberation when algebraic notation as we know today was discovered and used extensively by **Francois Viete (1540–1603)**, which is one of the reasons his work is a landmark in the development of algebra. One should also not forget that the development of notation was an evolving thing and continues right upto the present day. Even in Riemann's time $X^2, X^3, \ldots$ were often written as $XX, XXX, \ldots$. It took longer to have the fractional powers $X^{1/2}, X^{1/3}, \ldots X^{1/n}, \ldots$ replace $\sqrt{X}, \sqrt[3]{X}, \ldots \sqrt[n]{X}, \ldots$.

There was another obstacle faced by the mathematicians even upto the 16th century, namely the unwillingness to use negative numbers, even though the Hindus had successfully introduced them centuries earlier. Thus, Al–Khwarizmi , with the convention of writing only positive quantities, had to distinguish between different types of quadratic equations such as

$$X^2 = BX + C, \qquad X^2 + BX = C, \qquad X^2 + C = BX$$

Moreover these would have to be referred to in descriptive form as

> square equals thing and number
> square and thing equal number
> square and number equal thing

Al–Khwarizmi gave rules for solving these equations, presented examples to illustrate his rules, and essentially also giving demonstrations of the rules through these examples as well as geometrically.

TABIT BEN QURRA

Arabic mathematics continued to flourish after Al-Khwarizmi, and some of the notable names are **Tabit ben Qurra** (836–931) and **Omar Khayyam** (see [vW2], pp. 3–31) for a very detailed account of the works of these people). Tabit ben Qurra worked in astronomy, studied quadratic equations and their solutions, and also had original contributions to number theory. In the last he made a study of *amicable numbers*. Two positive numbers m and n are called amicable if each is the sum of the proper divisors of the other. Thus 284 and 220 are amicable. To see this we simply list the proper divisors of these two numbers.

$$284 : \text{ proper divisors } 1, 2, 4, 71, 142 : \text{ sum } = 220$$

$$220 : \text{ proper divisors } 1, 2, 4, 5, 10, 11, 20, 22, 44, 55, 110 : \text{ sum } = 284$$

Tabit's rule for amicable numbers is as follows. If

$$p = 3.2^{n-1} - 1, \qquad q = 3.2^n - 1, \qquad r = 9.2^{2n-1} - 1$$

are primes, then

$$M = 2^n pq \text{ and } N = 2^n r$$

are amicable. If we take $n = 2$, then we get $p = 5, q = 11, r = 71$ which are all primes, and then $M = 220$ and $N = 284$. The study of amicable numbers was continued by Fermat, Descartes, and Euler, among others.

OMAR KHAYYAM

Omar Khayyam (c. 1150–) is of course very well known through the translation of his poem the *Ruba'yat*. He was a poet, philosopher, astronomer, and mathematician who lived in the 11th century. In algebra his work was devoted to solving quadratic and cubic equations through geometrical constructions. However he was not the first to do so, although he systematically discussed many cases. It must be mentioned that the solutions are obtained as the coordinates of the points where certain curves meet, such as a circle and a parabola, or a circle and hyperbola, or a parabola and a hyperbola. Such methods were already known to the Greeks, and in any case, there is no possibility of getting any numerical estimates of the solutions by these geometrical constructions. See [vW2] for further details.

6

BEGINNINGS OF ALGEBRA IN EUROPE

It was only after the eleventh century or so that there was a reawakening of interest in mathematics in Europe. This must of course have been due to the various contacts with the middle and far east established through traders, who by themselves and with the help of interested scientific amateurs, played a vital role in the introduction of the results and the principal works of Arabic and Hindu mathematicians to European mathematicians. By about the 11th century methods of solving quadratic equations were well understood in Europe, thanks to the study of the Arabic and Hindu texts. There must also have been some awareness that some of the irrationalities encountered in Book X of Euclid's *Elements* such as

$$\sqrt{\frac{5+\sqrt{13}}{2}} - \sqrt{\frac{5-\sqrt{13}}{2}}, \quad \sqrt{\frac{\sqrt{3}+\sqrt{2}}{2}} - \sqrt{\frac{\sqrt{3}-\sqrt{2}}{2}}$$

led to equations of degree higher than 2. Thus

$$X = \sqrt{1+\sqrt{2}}$$

satisfies the equation

$$X^4 = 2X^2 + 1$$

As another example one might mention the fact that the problem of **trisecting an angle** led to **cubic equations**, although one cannot be sure to what extent this had been known to mathematicians in ancient times. If one remembers the trigonometric identity

$$\cos 3A = 4(\cos A)^3 - 3\cos A$$

then, as $\cos 60° = \frac{1}{2}$,

$$X = \cos 20°$$

is a solution of the cubic equation

$$8X^3 = 6X + 1$$

It is reasonably certain that people were able to solve some special cases of cubic and biquadratic equations, and that they were slowly becoming aware that a general method for solving these equations, similar to the one for solving a quadratic equation, was not available. It may therefore be said with some confidence that a variety of sources and demands had slowly but inevitably led to the following general question at the beginning of the 12th century, which is also the beginning of our story.

TO SOLVE CUBIC EQUATIONS LIKE

$$X^3 = AX + B, \qquad X^3 + AX = B$$

The first figure in this story is **Leonardo of Pisa**, commonly known as **Fibonacci**. We turn to him next.

FIBONACCI (=LEONARDO OF PISA)

Fibonacci (1180—1240) is short for *Filio Bonacci* which was how *Leonardo of Pisa* called himself, *a member of the Bonacci Family.* His most famous works are

1. Liber Abbaci (1202, revised 1208)
2. Flos (1225)
3. Liber Quadratorum

About 1225 Frederick II held court at Pisa, and on that occasion Fibonacci was presented to him, and it appears that *Flos* was presented to the Emperor then. On that occasion the mathematician *John of Palermo* presented a number of problems that Fibonacci solved. Two of these are mentioned in *Flos*.

Find a number X such that $X^2 + 5$ and $X^2 - 5$ are both squares

---

The solution

$$X = 3\frac{5}{12}, \qquad X^2 + 5 = \left(4\frac{1}{12}\right)^2, \qquad X^2 - 5 = \left(2\frac{7}{12}\right)^2$$

is presented in *Flos* but discussed in more detail in *Liber Quadratorum*.

The other problem is more startling. It is

Find a number X such that $X^3 + 2X^2 + 10X = 20$

In *Flos* Fibonacci proves that in the second problem, *X cannot be any of the irrationalities discussed in Euclid's Book X of his Elements, such as* $\sqrt{a + \sqrt{b}}$. He moreover presents an approximate solution in sexagesimal form as

$$X = 1; 22, 7, 42, 33, 4, 40$$

which is of course

$$X = 1 + \frac{22}{60} + \frac{7}{60^2} + \frac{42}{60^3} + \frac{33}{60^4} + \frac{4}{60^5} + \frac{40}{60^6}$$

In decimal notation this comes to

$$X \approx 1.368808106$$

On substituting this in the expression $Y = X^3 + 2X^2 + 10X$ we get

$$Y \approx 19.99999995$$

If we write

$$X = 1.368808110$$

we get

$$Y \approx 20.00000005$$

Thus

$$1.368808106 < X < 1.368808110$$

We cannot be adequately surprised by this. It is not just how accurate Fibonacci was that is remarkable; rather it is the insight into the structure of numbers and the process of approximating numbers by rational fractions revealed by this problem that is truly surprising. The conviction that for solving even simple looking equations one has to go beyond the irrationalities of Euclid coming out of his geometric constructions is something that was very much ahead of his time [vW2].

Here is a sample of some of the formulae Fibonacci obtained by manipulating the irrationalities.

$$\sqrt{4 + \sqrt{7}} + \sqrt{4 - \sqrt{7}} = \sqrt{14}$$
$$4 + \sqrt[4]{10} = \sqrt{16 + \sqrt{10} + 8\sqrt[4]{10}}$$

Fibonacci of course is most famous for the sequence of numbers called **Fibonacci Numbers**, which is

$$0, 1, 1, 2, 3, 5, 8, 13, 21, \ldots$$

The rule of formation is this: we start with 0 and 1 and from then on, each number is the sum of its two immediate predecessors. It came as a solution to the following problem. It was originally formulated as the propagation of sheep but I shall take the liberty of formulating it as the propagation of mathematicians.

> How many mathematicians will we produce in a dozen years if each mathematician produces one every year, who, from the second year on, is productive?

One can write down an explicit formula for the Nth Fibonacci number. On the basis of this formula one can show that if X_N is the number of mathematicians available after N years, then

$$X_N \approx (0.4472)(1.61803)^N$$

Thus the mathematicians increase like compounded capital at 60% interest!! Fibonacci numbers thus grow very fast with N, indeed in geometric progression. This is often called **exponential growth**. They remained as curiosities till in the 1960's they were found to be crucial in certain studies in mathematical logic. I remember an old cartoon in the *New Yorker* magazine showing two gentlemen walking down a street in which a certain house has the number 1123581321, and one of them is saying to the other "I suppose that is the house of old Fibonacci".

FORMULA FOR FIBONACCI NUMBERS

The problem is to determine explicitly the formula for the sequence of whole numbers defined by the rules

$$X_N = X_{N-1} + X_{N-2} \qquad X_0 = 0, X_1 = 1$$

It can be shown that

$$X_N = \frac{1}{\sqrt{5}}\left(\frac{1+\sqrt{5}}{2}\right)^N - \frac{1}{\sqrt{5}}\left(\frac{1-\sqrt{5}}{2}\right)^N$$

The absolute value of the second term is *strictly less than* $1/2$ *for all* N and so we have the alternative expression

$$X_N = \left\{\frac{1}{\sqrt{5}}\left(\frac{1+\sqrt{5}}{2}\right)^N\right\}$$

where $\{\ldots\}$ means *the integer nearest to*.... For $N = 7$ the actual value is 13 while the formula, before taking the integer nearest to it, gives 12.99, which is quite good!

Note that keeping the same recursion relation we may take other values for X_0 and X_1. the formula for the sequence will then change. We shall see examples of this in the exercises below.

NOTES AND EXERCISES

1. Let a, b be positive numbers with $a^2 > b$. Prove that

    $$x = \pm\sqrt{\frac{a + \sqrt{b}}{2}} \pm \sqrt{\frac{a - \sqrt{b}}{2}}$$

    are the roots of the equation
    $$x^4 - 2ax^2 + b = 0$$

2. Verify the following geometric construction of an irrationality like the one in the previous exercise. Let AB be the diameter of a semicircle of center F and radius 1 and let AB be extended to C so that BC is equal to the radius. Let CD be the tangent to the semicircle and let E be the midpoint of the arc BD. Then the length CE is given by

    $$\sqrt{5 - 2\sqrt{3}} = \sqrt{\frac{5 + \sqrt{13}}{2}} - \sqrt{\frac{5 - \sqrt{13}}{2}}.$$

    (Hint : Verify that the angle $< BCD$ is $30°$ and hence that the angle $< EFB$ is also $30°$. Hence deduce using trigonometry that BE^2 is $2 - \sqrt{3}$ and use the theorem of Apollonius in the form $CE^2 = 2BE^2 + 1$.)

3. Verify the trigonometric identity

    $$\cos 3\theta = 4\cos^3 \theta - 3\cos\theta$$

    and hence that $x = \cos 20°$ satisfies

    $$8x^3 - 6x - 1 = 0.$$

4. Assuming the explicit formula for the Fibonacci numbers deduce the two ways of describing the solution, namely that X_N is the integer nearest to

    $$\frac{1}{\sqrt{5}} \left(\frac{1 + \sqrt{5}}{2} \right)^N$$

    as well as

    $$X_N \approx (0.4472)(1.61803)^N$$

The following is a general method for finding the expression for the Fibonacci numbers as well as a number of other similar problems. Let (X_N) be a sequence of numbers which are defined by the *recursion relation*

$$X_N = pX_{N-1} + qX_{N-2}, \qquad X_0, X_1 \text{ arbitrary}$$

It is then clear that once X_0 and X_1 are given, one can calculate using this relation the values of $X_3, X_4, \ldots$ recursively (hence the name recursion relation). Here p and q are given numbers. For the Fibonacci sequence, $p = q = 1$ and $X_0 = 0, X_1 = 1$.

5. (a) Prove that if we take a *trial solution* of the form $X_N = x^N$, then x must satisfy the equation

$$x^2 = px + q$$

(b) Let a, b be the two roots of this quadratic equation. Show that for any constants A, B, the sequence

$$X_N = Aa^N + Bb^N$$

satisfies the recursion relation. Choose the constants so that X_0 and X_1 have given values.

(c) Deduce from the above the formula for the general term of the Fibonacci sequence discussed in the text. For the Fibonacci sequence with $X_0 = 1, X_1 = 3$, show that

$$X_N = \left(\frac{1 + \sqrt{5}}{2}\right)^{N+1} + \left(\frac{1 - \sqrt{5}}{2}\right)^{N+1}$$

(d) Find a formula for X_N where

$$X_N = 3X_{N-1} + X_{N-2}, \qquad X_0 = 0, X_1 = 1$$

FRA LUCA PACIOLI (1445—1514)

His main work is *Summa de arithmetica, geometrica, proportioni e proportion-alità*. Unlike most treatises of those times it was written in Italian rather than Latin and was published in Venice in 1494. It was extremely influential, especially because of the great stature of the author. At the end of this treatise Luca Pacioli wrote that the method of solving the cubic equations was not included in the book and mentioned this problem alongside the famous unsolved classical problem of squaring the circle. This led many to believe that the problem of solving cubic equations was hopeless.

Luca Pacioli already had a better notation for algebraic expressions than his predecessors. Thus roots were denoted by R (R for *Radix* or *Radical*, a terminology that is still used extensively). The square root was denoted by $R2$, the cube root by $R3$, the fourth root by RR (Radix Radix), and so on. The modern notations

$$\sqrt{\quad}, \quad \sqrt[3]{\quad}, \quad \sqrt[4]{\quad}$$

etc are of course derived from this. If the root was to be applied to an entire expression he wrote V before the root sign to indicate this ($V = U = Universale$). Plus and minus signs were denoted by p and m. Thus

$$RV\,40mR320$$

meant

$$\sqrt{40 - \sqrt{320}}$$

The unknown (the (in)famous X of high school mathematics!) was called *co (cosa)*, its square *ce (censo)*, its cube *cu (cubo)*, the fourth power *ce.ce (censo censo)*. If there appeared a second unknown quantity (the Y of high school algebra) it was called *quantità*. Thus the road to better notation was already started by Luca Pacioli, a journey that was to be completed, as we mentioned earlier, by Francois Viete[vW2].

7

THE CUBIC AND BIQUADRATIC EQUATIONS

SCIPIONE DEL FERRO

There was one man who was not deterred by Fra Luca Pacioli's comparison (implied at least) of the problem of solving cubic equations with the well known problem of antiquity, the problem of squaring the circle. That man was **Scipione del Ferro** (1465—1526) who was a Professor of mathematics for thirty years at the University of Bologna until his death in 1526. He solved the equation

$$X^3 + PX = Q$$

and communicated the solution to his son-in-law and successor on the faculty, **Annibale della Nave** and to his disciple **Antonio Maria Fior**. He did not publish his discoveries and so the evidence for his having solved the equation is indirect, but nevertheless convincing. It appears likely that he also solved the other forms of the equations (which, as I have explained earlier arise because of the insistence that no negative numbers appear and only positive solutions are sought after), namely

$$X^3 = PX + Q, \qquad X^3 + Q = PX$$

Eventually the formula for the solution of the cubic equation would become known as **Cardano's formula** although Cardano did not discover it. As we shall see later, Tartaglia rediscovered it in his contest with Fior (del Ferro's disciple) although he (Tartaglia) appears to have known that there was a formula in the possession of Fior. Cardano and his disciple Ferrari traveled from Milano to Bologna in 1543 and examined the personal notes of Scipione del Ferro with the permission of Annabelle della Nave. This examination convinced them that Scipione del Ferro did indeed possess the formula. Thus it is Scipione and not Tartaglia or Cardano who should be rightly regarded as the discoverer of this formula.

By any standards, Scipione del Ferro's achievement is remarkable. Unaided he had discovered the solution to a problem that had eluded many of his distinguished predecessors. The knowledge of his discovery slowly but inevitably set in motion the birth of Modern Algebra.

NICCOLÒ TARTAGLIA

Niccolò Fontana (Tartaglia) (c. 1500—1557) is, alongside **Gerolamo Cardano** and **Luigi Ferrari**, one of the principal figures in the story of the cubic and biquadratic equations. He was born in *Brescia* in 1499 or 1500 and was wounded in the larynx in a childhood accident so that he had a speech impediment. Hence he got his nickname *Tartaglia* which means *stutterer*. He went to Venice in 1534 where he was a mathematics teacher and had many contacts with the military people such as artillerymen and engineers there. This stimulated him scientifically and he thought about problems of ballistics. Thus he was a scientific forerunner of Galileo (nowadays he would have been regarded as a *Consultant*). He published two books:

1. *La nuova Scientia (The new Science)*

2. *Quesiti e Inventioni Diverse (Problems and various Inventions)*

These books were written in Italian rather than Latin perhaps reflecting his status as a figure with his feet in both the academic and the real worlds. In personal interactions he was reputed to be difficult.

The story of Tartaglia's role in the solution of the cubic equation is well known and easily told. In those days, problem-solving duels, where the stakes were considerable sums of money, were quite common in the academic world. They were a device either to establish one's preeminence, or destroy an already established character. Antonio Maria Fior, the disciple of the late Scipione del Ferro, who had in his possession the method that del Ferro had discovered for solving the cubic equation (and had given him before he died)

$$X^3 + PX = Q$$

issued such a challenge in 1535. Fior was a rather mediocre mathematician and sought to become established using his teacher's discoveries which he alone had access to. There were 30 problems in this contest that Fior proposed, with the stipulation that the loser should pay for a banquet for a group of 30 people consisting of Fior and his friends. Tartaglia accepted the challenge but discovered to his dismay that *all 30 problems depended on solving the above cubic equation for various values of P and Q!* Perhaps the prospect of personal and financial humiliation stimulated him, or perhaps he had an inexplicable stroke of inspiration; whatever the reason, Tartaglia discovered, almost when the actual public contest was about to begin, between February 12 and 13, the formula for the solution of the above equation. It is almost certain he knew that Fior had a formula for solving this equation in his possession and so there were no psychological barriers to his struggles. There are many well-documented instances in Science, where the mere knowledge that someone else has solved a problem makes it easy for others to find a solution themselves. In any case, much to the surprise of Fior, Tartaglia solved all 30 of Fior's problems, while Fior could not do any of Tartaglia's problems. To make the humiliation complete Tartaglia disdained the obligatory banquet by the loser Fior.

The story of the contest became widely known, making Tartaglia a much feared and admired teacher. At the same time, knowledge slowly spread in learned circles that the problem of the cubic equation had a solution and that Tartaglia had it in his possession.

As we shall see later when we discuss the life of Cardano, the method of solving the above cubic passed from Tartaglia to Cardano; indeed, Cardano succeeded in 1539 persuading Tartaglia to give him the method, with the promise that he (Cardano) will not publish it as his own. Cardano did keep his promise for four years, but learnt in 1543 that Scipione del Ferro had already discovered the method, and so published it, along with some of the additional discoveries he made, in his famous treatise *Ars Magna*. This infuriated Tartaglia and led to a lot of unpleasantness between Tartaglia, Cardano, and Ferrari, the disciple of Cardano. Ferrari, as we shall see later, discovered how to reduce the solution of a *biquadratic equation* to that of a cubic.

In some sense Tartaglia is a tragic figure because his rediscovery of Scipione del Ferro's solution did not get him enough credit either during his lifetime or in the eyes of posterity. To some extent his secretive and difficult personality was responsible for his difficulties during his lifetime. Also he had the misfortune to clash head to head with the two mathematicians in the whole world who were superior to him in his field. The fact that he did not leave any manuscript or book in which this problem was treated added to the difficulties of later-day researchers in highlighting his role. Perhaps the most interesting lesson from this story is not that this or that person did or did not receive full credit for what he did, but that the power of ideas is irresistible. Once Scipione del Ferro made the magical discovery, no power on earth could stop it from slowly spreading. Ideas cannot be bottled and stored like wine or patented like a secret method of making 100 pancakes in 5 minutes.

One must not imagine that the secretiveness of Tartaglia and the fiercely possessive stance he assumed towards the dissemination of his discovery were special to those days. They persist to this day and act as real obstacles to the spread of knowledge. There are many instances of scientists and mathematicians who make great discoveries and are then dismayed to find others working on them, refining them, and improving them. Fortunately such instances are not too common. The great majority of workers in Science recognize the universal and public character of scientific discoveries and participate willingly in what they know to be a collective endeavour to advance the frontiers of what we know.

GEROLAMO CARDANO

It is not given to many to inaugurate an era in human affairs. **Gerolamo Cardano** (1501—1576), Renaissance scholar, physician, mathematician, a person with an encyclopedic and universal cast of mind, is one of these. Born of modest circumstances, he rose to fame and importance entirely by his own efforts. His works were widely read already in his own time, and he was considered, along with Vesalius, as the foremost physician of his epoch. In mathematics, which he pursued as a hobby, he is famous as the author of *Ars Magna (The Great Art)*, a book that marks the beginning of the subject of *Modern Algebra*. His book *Liber de Ludo Aleae (The book on games of chance)* is considered by many to be the first serious effort to codify mathematically the probabilities that arise in card and dice games, and makes him a substantial figure in the history of the *Theory of Probability*, ranking along with **Pierre de Fermat** and **Blaise Pascal**. He wrote voluminously and on every conceivable topic. His collected works, which first appeared in *Lyons* in 1663, runs to ten huge volumes. One of the most famous and most widely read of his books is his autobiography *De Propria Vita* which is a remarkable document of wonderful human interest; it succeeds as no other book had done upto that time, in bringing out the picture of a man with many faults and troubles, but still able to rise above them to a life of unsurpassed achievement. The great mathematician and philosopher **Leibniz** offered perhaps the most suitable epitaph on Cardano when he wrote: "Cardano was a great man with all his faults; without them he would have been incomparable".

Cardano was born in 1501 in Milano. His father **Fazio Cardano** was a lawyer there but was deeply interested in the medical sciences and in mathematics; at one time he was giving systematic public lectures on Geometry in the University of Pavia and later on in Milano under the auspices of a certain charitable foundation. Fazio was well known and was often consulted on geometrical questions by Leonardo Da Vinci himself. Gerolamo went to the University of Pavia to study medicine and finished his medical studies successfully in 1526. But his difficult personality must have offended many; he was denied the privilege of entering the College of physicians and practicing in Milano on the flimsy excuse that he was of illegitimate birth. So, for a few years he practiced medicine in a small town called Sacco, near Padua. But, tiring of this, he came back to Milano and was in great difficulties for some time, even being forced to spend some in the poorhouse with his wife and son. Luckily for him his career took a turn for the better. His deep interest in scientific questions and his capacity to explain them with lucidity to the lay person, impressed many of the nobles who were influential, and he was appointed as the public lecturer in geometry, the same post that was earlier occupied by his father. He wrote two mathematical books during this time on elementary aspects of Algebra which were widely read.

The College of physicians in Milano was still adamant in refusing membership to him, and Cardano, trying to reverse this situation, published a book containing a withering assault on the practices of medicine of that time. The book created considerable interest in the public and dismay in the College, and the College, perhaps thinking that he may be easier to control from within as a member rather than as an outsider, admitted him finally. He rapidly became one of the greatest of the practicing physicians of his time. He was sought after not only in Italy but also from other countries. One of his most notable accomplishments was to effect a complete cure to the severe asthmatic and allergic difficulties of John Hamilton, the Archbishop of Scotland.

His children however did not bring him peace and happiness. His eldest son had a stormy married life and he poisoned his wife, a crime for which he was tried, found guilty, and executed in 1560. Cardano never recovered from this blow and left Milano for the University of Bologna where he took over the position of the Professor of Medicine in 1562. But on October 6, 1570, he was suddenly arrested and imprisoned as a heretic. However, in consideration of his great eminence and the testimony of many of his friends and disciples, he was released after a few months, and allowed to live without surveillance in his own house. Later he moved to Rome with the help of one of his friends, **Rodolfo Silvestri** and surprisingly, was invited to be a member of the College of physicians there. He lived the remainder of his life in peace and quiet, writing books and articles till the end. He died peacefully on September 20, 1576.

This is just a hurried glimpse into the life of one of the great figures of Renaissance scholarship and creativity. His book *Ars Magna* ranks with the books of Copernicus and Vesalius as one of the most important books of the 16th century. In a very real sense he was a figure who turned away from the dark and secretive practices of the middle ages and looked forward into the era of enlightenment, who believed in the power of ideas and the necessity of their dissemination, and who laid the foundation of one of the great disciplines in mathematics, namely *Modern Algebra*. To get a real perspective on his life and accomplishments one should refer to the book *Cardano: The Gambling Scholar* by **Oystein Ore** (Princeton University Press, 1953), as well as read some of his books, especially *Ars Magna* and *De Propria Vita*.

LUIGI FERRARI

The last of the dramatis personae in the story of the cubic and biquadratic equations is **Luigi (Lodovico) Ferrari** (1522—1565). He came to Cardano as a youth on November 30, 1536, and Cardano has described that day with a lot of enthusiasm in his writings. Ferrari was fourteen at that time and he came from Bologna to work in the household of Cardano. But Cardano realised very soon that he had an exceptional servant and from then on treated him first like a disciple, and then as a collaborator. For his part, Cardano had no more loyal friend than Ferrari. Indeed, in the dispute with Tartaglia, when the latter referred to Ferrari as "Cardano's creature", Ferrari did not take exception to this, but rather felt proud.

Ferrari was a genius but had an explosive temper. At times even Cardano was afraid to speak to him. Once when he was seventeen he returned from a brawl with all the fingers in his right hand missing. But he learned fast and by the time he was twenty he was a public lecturer in mathematics, his preferred subject. His lectures were very popular and he became very famous in his life time.

He learnt the solution of the cubic equation from Cardano and soon made his own contribution when he discovered the method of solving the *biquadratic equation*, namely the equation of degree 4. His method was to reduce it to the solution of an auxiliary cubic equation which would then be solved by Cardano's method. In some sense this was a watershed event in the theory of solution of equations. The equation of the fifth degree resisted all methods of solution for a century and more for reasons that nobody understood for a long time. It was only with the work of the nineteenth century masters **Abel, Ruffini, and Galois** that mathematicians understood finally that a solution of the fifth degree equation along the lines of the cubic and biquadratic equations was impossible, thereby inaugurating the modern theory of groups.

When Tartaglia, furious that Cardano had broken his solemn promise when he published his ideas (with the most proper acknowledgments as we shall see later) in the *Ars Magna*, challenged Cardano to a public dispute, it was Ferrari who answered the challenge and defeated Tartaglia in one of the most famous of such encounters from the sixteenth century. Earlier, it was Ferrari who accompanied Cardano in the journey to Bologna to discover whether Scipione del Ferro had already discovered the solution of the cubic equation before Tartaglia. He was thus at the side of Cardano in the most important scientific occasions of Cardano's life. His death, in 1565, was a profound loss to Cardano. Cardano wrote an account of Ferrari's life and achievements and acknowledged that he had no more loyal friend and coworker.

THE DISPUTE

This is a brief sketch of the dispute between Cardano and Ferrari on one side and Tartaglia on the other. As we now know, Scipione del Ferro had already discovered before his death in 1526 the method of solving at least one of the three forms of the cubic equation. His secret passed on to his succeessor in the department at Bologna, namely his son-in-law Annibale della Nave, and his pupil, Antonio Maria Fior. In order to establish himself, Fior then issued a public challenge with a list of 30 problems all of which led to the solution of the cubic. We must remember that Fior probably thought nothing of the fact that the weapon that he was using was not devised by him. No one at that time knew his source, but even had it be known, the standards of behaviour of the times might not have resulted in any serious condemnation of what Fior did. In any case, the challenge was issued in 1535 and the opponent was Tartaglia who submitted a list of counter problems. By what must have been a great effort Tartaglia discovered the way to solve the form of the cubic equation on which the problems of Fior rested and won the dispute decisively. The strategy of Fior had backfired.

Of course once the challenge and its results became public it must have been realized that Tartaglia had the solution of at least one form of the cubic equation. Cardano, who was writing a book on Algebra at that time, also must have realized this and wanted to include this method in his book. So in 1539 he wrote to Tartaglia suggesting a meeting to discuss this. At first Tartaglia was insulting and contemptuous of Cardano and refused to meet him. But after a while he agreed, because Cardano offered to introduce him to the Governor of Milan which would give Tartaglia an opportunity to display his military inventions before the Governor and perhaps sell some of them to him. So Tartaglia accepted the invitation of Cardano and the two met in Cardano's house on March 25, 1539. Tartaglia subsequently reported on this meeting in the form of a dramatic dialogue in his account; we must note that Cardano left no account of it, nor did Ferrari, who was already in the Cardano household at that time, and must have been present during some parts at least of the conference.

At first Tartaglia was adamant in refusing to divulge his discovery because his suspicious mind felt that Cardano would use it without proper acknowledgement to his own glory. But suddenly he relented but on one condition: he wanted Cardano to swear under oath that he would not publish this secret. Cardano did this and received the secret from Tartaglia. Tartaglia then left town in a hurry, without even seeing the Governor as arranged by Cardano. For some years after this Cardano kept his promise. But slowly he must have realized that Fior would not have issued the challenge unless he knew how to solve the equation, and that meant, as Fior was not a sharp mathematician, that del Ferro must have done it and passed it on to his pupil. So in 1543 he journeyed to Bologna with Ferrari to settle this. In Bologna he went to della Nave's house and got his permission to look through the papers of del Ferro. This study convinced both Cardano and Ferrari that del Ferro had solved the equation before his death. This circumstance altered the situation of the oath in Cardano's mind. He must have felt that as the discovery was not Tartaglia's but already del Ferro's, the oath was no longer binding on him. It is possible to argue this point, but the fact remains that after their return from Bologna, Cardano lost no time in preparing and publishing his great masterpiece *Ars Magna*. Of course his treatment of priorities was most scrupulous; for the original discovery he took no credit but attributed it to del Ferro and pointed out that Tartaglia also discovered it independently. Even by contemporary standards the acknowledgements of Cardano are most proper.

There was another circumstance which must have played a big role in the decision of Cardano to go public. He had not been sitting still after getting Tartaglia's secret; he began exploring the other cases of the cubic equation and succeeded in verifying that the formula of Tartaglia could be given such an expression that it applied to all the forms of the cubic. To his surprise he found that this led him to imaginary numbers which he had to manipulate as if they obeyed all the properties of ordinary numbers (such calculations he calls *truly sophisticated* in his book). He essentially discovered the relationship between roots and coefficients, and the fact that the general cubic had either a unique real root, in which case Tartaglia's formula led to it once taking cube roots was permitted, or else it had three real roots, in which case the more general formula he had worked out described all three of them, but in a form in which imaginary numbers made their appearance in an essential manner. He called this the *irreducible case*. If we remember the reluctance of mathematicians to use even negative numbers at that time, the researches of Cardano must be regarded as incredibly pioneering. This was not all; Ferrari must have participated in many of these discoveries at least after Cardano made them, and he added his own, namely a decisive method of solving the equation of degree 4. Thus Cardano and Ferrari were sitting on a huge pile of original and path-breaking discoveries and getting frustrated by their inability to make thempublic because of the oath.

Once *Ars Magna* appeared in print, Tartaglia wasted no time in responding. He gave an account of his secret meeting with Cardano, the story of the oath, and castigated Cardano in the most scathing terms as a breaker of a solemn promise. The fact that Cardano took no credit for the original discovery and attributed it to del Ferro and Tartaglia did not mollify him. Cardano did not respond to this but Ferrari did in the form of a public challenge again. The dispute took place on August 10, 1548, in Milano. Again only Tartaglia has left an account of what happened on this occasion. Tartaglia's account appears self-serving and is not able to hide the fact that he was defeated. In any case he left the dispute before its end, a circumstance that almost certainly must have led to his being named the loser. But what happened after the dispute is beyond any doubt; Tartaglia went back to Venice because the offers of appointments did not come through (it is difficult to avoid the inference that these offers must have been contingent on his winning the dispute and so must have been withdrawn after his defeat), while Ferrari went on to a very distinguished and public career.

How are we to judge these events? Or, as is more appropriate, should we judge at all? There are mitigating circumstances on both sides, which is reflected in the fact that historical judgements have fluctuated back and forth. At first historians sided with Cardano and Ferrari, but later on, a revisionist view held that Tartaglia was the wronged party. It is likely that the truth (if there is a truth in such cases) is in between somewhere, and that as in all human affairs of any importance, both sides are correct. But in one aspect there is no disagreement; Tartaglia, through his penchant for secrecy, represents the middle ages, while Cardano, with his generous views about authorship and sharing of knowledge, represents the modern view. Mathematics was very fortunate that a person with the great vision and world-view like Cardano was able to get his hands on the original discovery and was then able to build a wonderful structure on top of it that led to the beginning of Algebra as we know today.

For a much more detailed account with greater perspective of all these things, see [G].

SOLUTION OF THE CUBIC
AND BIQUADRATIC EQUATIONS

8

SOLUTION OF THE CUBIC EQUATION

REVIEW OF QUADRATIC EQUATIONS

Before taking up the cubic equations let us review the theory of quadratic equations.

$$
\begin{aligned}
\text{EQUATION} \quad &: \quad X^2 + PX + Q = O \\
\text{DISCRIMINANT} \quad &: \quad \Delta = P^2 - 4Q \\
\text{ROOTS} \quad &: \quad X = -\frac{P}{2} \pm \frac{1}{2}\sqrt{\Delta}
\end{aligned}
$$

Although these formulae involve square roots, sums and products of roots can be computed from the coefficients directly. In fact any polynomial function of the roots which is symmetric in the two roots (as the sum and product are) can be evaluated in terms of the coefficients alone. We shall see how this can be proved in the exercises below.

$$
\begin{aligned}
\text{EQUATION} \quad &: \quad X^2 + PX + Q = O \\
\text{ROOTS} \quad &: \quad X_1, X_2 \\
X_1 + X_2 \quad &: \quad -P \\
X_1 X_2 \quad &: \quad Q
\end{aligned}
$$

This process can be reversed. Thus given the sum and products of roots the equation is unique and is written down immediately.

$$
\begin{aligned}
\text{SUM OF ROOTS} \quad &: \quad A \\
\text{PRODUCT OF ROOTS} \quad &: \quad B \\
\text{EQUATION} \quad &: \quad X^2 - AX + B = 0
\end{aligned}
$$

This solves an ancient problem: how to divide a number A into two parts whose product is to be another number B. *This result will be used in the solution of the cubic equation.*

One can interpret the formula for the roots of a quadratic equation in two ways. The first is to view it as an expression in terms of the coefficients of the equation, that, when substituted for X in the equation and subjected to algebraic operations, yields 0 identically in the coefficients. The $\pm$ sign in the root means that either of the two expressions will have this property. The second interpretation is that if we give specific values to the coefficients of the equation, the formula will yield a number that will satisfy the equation. However the formula for the roots contains a square root and it is not clear what its meaning is if the number Δ turned out to be negative. It was not till much later, in the eighteenth and nineteenth centuries, when *complex numbers* were invented, that one knew how to interpret the formula for the roots when $\Delta < 0$. But, as we shall see later, the beginning of the solution to this difficulty was already made by Cardano.

NOTES AND EXERCISES

1. Let a, b be the roots of the quadratic equation $X^2 - AX + B = 0$. Find expressions for $a^2 + b^2$ and $a^3 + b^3$ in terms of A and B. (Hint : Write $a^2 + b^2$ as $(a+b)^2 - 2ab$ and use the expansion for $(a+b)^3$ to find an analogous expression for $a^3 + b^3$.)

2. Let $S_r = a^r + b^r$ where a, b are as in the previous exercise. Prove that
$$S_1 S_{r-1} = S_r + B S_{r-2}, \qquad S_1 = A$$
Hence prove by induction on r that each S_r is a polynomial in A and B with integer coefficients. Obtain by this method the expressions for S_5 and S_6 in terms of A and B.

3. Suppose that P is a polynomial in a and b with integer coefficients which is *symmetric*, i. e., is unchanged when we interchange a and b. Show that P is a polynomial in A and B with integer coefficients also. (This is a very special case of *Newton's theorem* which asserts a similar result for equations of any degree. To prove this, look at the terms of the highest degree in P and observe that they are linear combinations with integer coefficients of terms $(ab)^k(a^r + b^r)$, and use the preceding exercise.)

4. There are some special equations of degree 4 which can be solved using the solution of quadratic equations. Prove that the equation
$$X^4 + PX^2 + Q = 0$$
has the solutions
$$X = \pm\sqrt{\frac{-P}{2} \pm \frac{1}{2}\sqrt{\Delta}}$$
Hence show that the solutions of $X^4 - 3X^2 + 1 = 0$ are
$$\pm\frac{\sqrt{6 \pm 2\sqrt{5}}}{2}$$
Find the approximate value to 6 places of these roots in the sexagesimal system.

5. For the equation

$$X^4 + PX^2 + Q = 0$$

find expressions in terms of P and Q for the following functions of the roots:

$$\sum_i a_i^2, \qquad \sum_i a_i^3, \qquad \sum_i a_i^4$$

SCIPIONE DEL FERRO'S FORMULA

We now take up the cubic equation. We shall see later that it is enough to consider the case when the coefficient of X^2 is 0. Thus the equation to be considered is

$$X^3 + PX = Q$$

Once again what is sought for is an expression in terms of the coefficients P, Q, that, when substituted for X will yield 0 identically in the coefficients. By analogy with the quadratic case the expression for the root is allowed to involve cube roots. Moreover, when the coefficients are fixed, this expression must make sense under certain conditions and yield a root of the equation. Scipione del Ferro came up with precisely one such formula.

When the equation is written in the above form it was implicitly assumed by the mathematicians of those times that *P and Q are positive numbers*. The expression that Scipione del Ferro obtained will then make sense as we shall see presently. It can be shown, using calculus for example, that in this case the cubic equation above has only one real root. It is really remarkable that the special case considered by Scipione del Ferro is such that there is a unique real root, so that one can speak of *the formula for the root*. The formula is then given in the following form.

$$
\begin{array}{rcl}
\text{EQUATION} & : & X^3 + PX = Q \\[2mm]
\text{DISCRIMINANT} & : & \Delta = \dfrac{Q^2}{4} + \dfrac{P^3}{27} \\[3mm]
\text{ROOT} & : & X = \sqrt[3]{\sqrt{\Delta} + \dfrac{Q}{2}} - \sqrt[3]{\sqrt{\Delta} - \dfrac{Q}{2}}
\end{array}
$$

There is no need to add that this is a very beautiful formula, even today. That Scipione del Ferro was able to derive it is indeed a great achievement for his time.

Equally remarkable is the fact that Tartaglia was able to discover it under pressure. As an illustration let us consider an example:

$$X^3 + 6X = 7$$

Here

$$P = 6, \qquad Q = 7$$

Hence

$$\Delta = (49/4) + 8 = (81/4), \quad \sqrt{\Delta} = 9/2$$

Hence del Ferro's formula gives

$$X = \sqrt[3]{(9/2) + (7/2)} - \sqrt[3]{(9/2) - (7/2)}$$
$$= \sqrt[3]{8} - \sqrt[3]{1}$$
$$= 2 - 1$$
$$= 1$$

However there is more in this than meets the eye. Let us consider another example, this time one from *Ars Magna* ([RW], p 99):

$$X^3 + 6X = 20$$

Here

$$P = 6, \qquad Q = 20$$

Hence

$$\Delta = 100 + 8 = 108, \quad \sqrt{\Delta} = 6\sqrt{3}$$

Hence del Ferro's formula gives

$$X = \sqrt[3]{6\sqrt{3} + 10} - \sqrt[3]{6\sqrt{3} - 10}$$

But a direct check reveals that

$$X = 2$$

satisfies the equation and so must be the root! In other words, we have the remarkable identity

$$2 = \sqrt[3]{6\sqrt{3} + 10} - \sqrt[3]{6\sqrt{3} - 10}$$

Try proving this directly!!

NOTES AND EXERCISES

1. For the equation

$$X^3 + 9X = 10$$

   show that $X = 1$ is a root, while $\Delta = 52 > 0$ and so there del Ferro's formula gives the identity

$$1 = \sqrt[3]{2\sqrt{13} + 5} - \sqrt[3]{2\sqrt{13} - 5}$$

2. Find the expression for the root of the cubic equation

$$X^3 + 3X = 6$$

   by del Ferro's formula.

PROOF OF DEL FERRO'S FORMULA

We shall now give a proof of the formula of del Ferro. Since we have the formula, it is just a question of substituting in the equation and verifying that we get 0, although in those times, with virtually no decent notation to speak of, this was a huge task. Nevertheless it is natural to feel dissatisfied with a proof that just *verifies* the formula. After all, del Ferro did not have the formula! So we shall make a compromise, and prove the formula assuming that the root has a form *qualitatively similar* to the expression to be proved. The proof will explain what we mean by this.

The starting point is the formula for computing the cube of a sum :

$$(A + B)^3 = A^3 + B^3 + 3AB(A + B)$$

This is perhaps not as familiar as the corresponding formula for $(A + B)^2$ and so let us indicate how this is obtained. We just start with the formula for the square and multiply it by $A + B$. Thus

$$\begin{aligned}
(A + B)^3 &= (A + B)(A + B)^2 \\
&= (A + B)(A^2 + 2AB + B^2) \\
&= A^3 + B^3 + 3A^2B + 3AB^2
\end{aligned}$$

This gives

$$(A+B)^3 = A^3 + 3A^2B + 3AB^2 + B^3$$
$$= A^3 + B^3 + 3AB(A+B)$$

The key observation now is that *the formula for X expresses it as the sum of 2 cube roots*. So let us make a *tentative assumption* that X has an expression *as a sum of two cube roots*, i.e.,

$$X = \sqrt[3]{U} + \sqrt[3]{V}$$

*where U and V are expressions to be determined later on.*Then, if we take

$$A = \sqrt[3]{U}, \qquad B = \sqrt[3]{V}$$

in the earlier formula we get

$$X^3 = U + V + 3\sqrt[3]{U}\sqrt[3]{V}(\sqrt[3]{U} + \sqrt[3]{V})$$
$$= [U + V] + 3\sqrt[3]{U}\sqrt[3]{V}X$$

So

$$X^3 + [-3\sqrt[3]{U}\sqrt[3]{V}]X = U + V$$

If this is to be the same as the equation for X we must take

$$U + V = Q, \qquad -3\sqrt[3]{U}\sqrt[3]{V} = P$$

The second relation has cube roots which are gotten rid of by cubing. Thus we get

$$U + V = Q, \qquad UV = -\frac{P^3}{27}$$

In other words, and this is the heart of the argument, *the problem of finding U and V is now seen as just the ancient problem of determining two quantities whose sum and product are known.* Therefore we can say that U and V are the roots of the *quadratic* equation in Y,

$$Y^2 - QY - \frac{P^3}{27} = 0$$

This is of course immediately solved, giving us the formula for U and V:

$$U, V = \frac{Q}{2} \pm \sqrt{\Delta}, \qquad \Delta = \frac{Q^2}{4} + \frac{P^3}{27}$$

Since

$$X = \sqrt[3]{U} + \sqrt[3]{V}$$

the above expressions for U, V, when substituted, give

$$X = \sqrt[3]{\frac{Q}{2} + \sqrt{\Delta}} + \sqrt[3]{\frac{Q}{2} - \sqrt{\Delta}}$$

Let us now examine whether this formula for the root of

$$X^3 + PX = Q, \qquad (P > 0, \quad Q > 0)$$

will make sense when P and Q are positive numbers. Since Δ is > 0, the first term makes sense. For the second term note that

$$\frac{Q}{2} < \sqrt{\Delta}$$

since this is equivalent to

$$\frac{Q^2}{4} < \frac{Q^2}{4} + \frac{P^3}{27}$$

So the expression inside the cube root sign is < 0. But then, as

$$\sqrt[3]{-W} = -\sqrt[3]{W}$$

we get, reversing the signs within the cube root sign and taking the minus sign outside the cube root,

$$X = \sqrt[3]{\sqrt{\Delta} + \frac{Q}{2}} - \sqrt[3]{\sqrt{\Delta} - \frac{Q}{2}}$$

which is Scipione del Ferro's formula!

Cardano describes the formula in words in his book *Ars Magna* as follows ([RW], p 98):

Cube one-third the coefficient of X; add to it the square of one-half the constant of the equation; and take the square root of the whole. You will duplicate this and to one of the two you add the number you have already squared and from the other you subtract the same. You will then have a *binomium* and its *apotome*. Then subtracting the cube root of the *apotome* from the cube root of the *binomium*, the remainder [or] that which is left is the value of X.

Thus, for the equation

$$X^3 + PX = Q$$

if we follow the rule we should proceed as follows. Cube

$$\frac{P}{3}$$

which is

$$\frac{P^3}{27}$$

Square

$$\frac{Q}{2}$$

which is

$$\frac{Q^2}{4}$$

The result is

$$\Delta = \frac{Q^2}{4} + \frac{P^3}{27}$$

The *binomium* and the *apotome* are, respectively,

$$\sqrt{\Delta} + \frac{Q}{2}, \qquad \sqrt{\Delta} - \frac{Q}{2}$$

and the value of X is

$$X = \sqrt[3]{\sqrt{\Delta} + \frac{Q}{2}} - \sqrt[3]{\sqrt{\Delta} - \frac{Q}{2}}$$

Cardano of course does not quite give the formula as we have given but illustrates the rule by applying it to many examples, as was the custom in his time to describe general rules without notation.

REMOVING THE X^2 TERM

We shall now go a little more into the theory of the general cubic equation by considering other forms of the equation. First of all let us not assume that the coefficient of the X^2 term is 0 and also not assume anything about the signs of the remaining coefficients. This is somewhat bold since we already know that in the theory of quadratic equations there is a problem when the discriminant is < 0.

Thus we shall write the general cubic equation as

$$X^3 + AX^2 + BX + C = 0$$

The first step is to reduce its solution to an equation *in which there is no term containing X^2*. The way to do it is to make a transformation from X to Y where

$$Y = X - t, \qquad X = Y + t$$

Here t is a constant which will be chosen presently. From the equation for X we obtain the equation for Y by substituting for X the expression $Y + t$. If we are then able to solve the equation for Y, the solutions of the equation for X are obtained by adding t to the solutions for Y. Thus

$$(Y + t)^3 + A(Y + t)^2 + B(Y + t) + C = 0$$

which simplifies to

$$Y^3 + (3t + A)Y^2 + (3t^2 + 2At + B)Y + (t^3 + At^2 + Bt + C) = 0$$

We can make the coefficient of Y^2 to vanish in excatly one way, namely by selecting the value $-A/3$ for t:

$$t = -\frac{A}{3}, \qquad X = Y - \frac{A}{3}$$

The equation for Y then becomes

$$Y^3 + B'Y + C' = 0$$

where

$$B' = -\frac{A^2}{3} + B$$
$$C' = \frac{2A^3}{27} - \frac{BA}{3} + C$$

It is not necessary to memorize this formula but only remember the method; in any given situation one can perform this calculation directly.

Let us try an example and consider the equation

$$X^3 = 6X^2 + 20$$

We write

$$X = Y + t$$

and get

$$Y^3 + 3tY^2 + 3t^2Y + t^3 = 6(Y^2 + 2tY + t^2) + 20$$

which, when we choose $t = 2$, becomes

$$Y^3 = 12Y + 36$$

This is not in the form of del Ferro, but let us be bold and try his formula for the root anyway:

$$\Delta = (18)^2 + (-4)^3 = 260$$

so that *the formula makes sense* and we can write

$$Y = \sqrt[3]{\sqrt{260} + 18} - \sqrt[3]{\sqrt{260} - 18}$$

from which we get, for X, the value

$$X = 2 + \sqrt[3]{\sqrt{260} + 18} - \sqrt[3]{\sqrt{260} - 18}$$

The moral of this calculation is that when we remove the coefficient of X^2 we may get an equation not covered by del Ferro's assumptions on the signs of the coefficients of X and the constant term. Nevertheless the formula will sometimes make sense and lead to a root. This is an important matter and we shall turn to it next.

SOLUTION IN THE GENERAL CASE

We shall now consider the equation

$$X^3 + PX = Q$$

where P and Q may be given values not necessarily positive. The first idea is to see what happens if we apply the formula of del Ferro; we had already done this in an example and come up with a result that makes sense. But this is not always going to be the case. Thus for the equation

$$X^3 - 6X = 2$$

the value of the discriminant is

$$\Delta = (2/2)^2 + (-6/3)^3 = 1 - 8 = -7$$

and the formula becomes

$$X = \sqrt[3]{1 + \sqrt{-7}} + \sqrt[3]{1 - \sqrt{-7}}$$

There is now the problem of having a number < 0 inside a square root sign. Cardano was aware of such cases and called them *truly sophisticated.*

There are now two questions, which are closely related, that arise:

> (i) Is the formula of del Ferro true identically in the coefficients?
>
> (ii) If we get negative numbers inside the square root, what is the meaning of the formula?

The first question is easy to answer. In fact, the method that we used earlier applies without change. Thus we suppose for X an expression of the form

$$X = \sqrt[3]{U} + \sqrt[3]{V}$$

where U and V are to be chosen properly. Then

$$X^3 = U + V + 3\sqrt[3]{U}\sqrt[3]{V}(\sqrt[3]{U} + \sqrt[3]{V})$$
$$= [U + V] + 3\sqrt[3]{U}\sqrt[3]{V}X$$

So

$$X^3 + [-3\sqrt[3]{U}\sqrt[3]{V}]X = U + V$$

If this is to be the same as the equation for X we must take

$$U + V = Q, \qquad -3\sqrt[3]{U}\sqrt[3]{V} = P$$

The second relation has cube roots which are gotten rid of by cubing. Thus we get

$$U + V = Q, \qquad UV = -\frac{P^3}{27}$$

Therefore we can say that U and V are the roots of the *quadratic* equation in Y,

$$Y^2 - QY - \frac{P^3}{27} = 0$$

This is of course immediately solved, giving us the formula for U and V:

$$U, V = \frac{Q}{2} \pm \sqrt{\Delta}, \qquad \Delta = \frac{Q^2}{4} + \frac{P^3}{27}$$

Since

$$X = \sqrt[3]{U} + \sqrt[3]{V}$$

the above expressions for U, V, when substituted, give

$$X = \sqrt[3]{\frac{Q}{2} + \sqrt{\Delta}} + \sqrt[3]{\frac{Q}{2} - \sqrt{\Delta}}$$

We shall call this *Cardano's formula*, due to the circumstance that it is valid irrespective of the signs of P and Q. It is formally the same as del Ferro's, except for the + sign in the second term which is explained by reversing the sign of the expression in the second cube root in del Ferro's formula. It is clear from the structure of the formula that when the number Δ is positive (or zero), one can extract its square root and then it does not matter what the sign of the expression inside the cube root is, because one can take the cube roots of both positive and negative numbers. So when the coefficients P and Q are such as to make $\Delta > 0$, the formula of Cardano makes sense and gives the value of X. The point here is that Δ is certainly > 0 if P and Q are both positive, but it *can be* > 0 even when P is < 0 because it is possible that the size of $Q^2/4$ can overcome the negative value of $P^3/27$.

The second question is more subtle. It has nothing to do with the fact that we are dealing with a cubic equation since it is already present in the quadratic case. To answer it effectively *it is necessary to create a new number system* in which square roots of negative real numbers make sense. This is the system of *complex numbers*. Cardano had a glimpse of this world but it did not become a reality till much later, after a century and more of efforts by mathematicians, starting with **Bombelli** (c. 1526–1573). It was finally **Gauss** (1777–1855) who introduced the complex number system in full rigour and studied its properties. We shall later on follow this route and see how natural it is to introduce these new numbers.

We shall now summarize the situation concerning the solution of the cubic equation:

(i) We can always take the equation in the form $X^3 + PX = Q$.

(ii) If Δ is > 0, then the equation has a unique real root which is given by del Ferro's formula.

(iii) If Δ is < 0, the equation has *three* roots, *all of them real*, and the formula will still represent them, but one has to base the interpretation on the theory of *complex numbers*!

Cardano called the last case *the irreducible case*. That the roots, which are all real, have to be expressed in terms of complex numbers is one of the subtle aspects of the theory of the cubic equation with real coefficients. Thus, for cubic equations with real coefficients, when there are three distinct real roots, it is not possible to express the roots in terms of cube roots and square roots involving only positive numbers inside the root signs.

It is of course perfectly possible that $\Delta = 0$. Note that when $\Delta = 0$, either P and Q are both nonzero, or both of them are zero. In the latter case the equation is $X^3 = 0$ for which $X = 0$ is the only solution, although we shall say that 0 is a *triple root*. If $\Delta = 0$ but P and Q are nonzero, then the equation has two distinct real roots, but one of them is a double root. To be sure we have not proved all these statements, especially the ones relating the number of real roots to the sign of Δ, nor have we defined the concept of a double (or multiple) root. We shall give a glimpse into the proofs by assuming a naive approach to real numbers.

NOTES AND EXERCISES

1. For the cubic equation
$$X^3 - 6X - 4 = 0$$
    show that the formula of Cardano gives
$$X = \sqrt[3]{2 + \sqrt{-4}} + \sqrt[3]{2 - \sqrt{-4}}$$
    whose interpretation will have to wait till we introduce complex numbers.

NATURE OF THE ROOTS : DISCUSSION

We shall now discuss in a somewhat informal manner the nature of the roots of a cubic equation

$$X^3 + PX = Q$$

where P and Q are "real" numbers. We put the word real inside quotation marks because we wish to emphasize that we do not wish to worry too much about the foundations of the real number system but to take a rather naive approach to it.

The first observation is that *there is always at least one real root.* To see this let us write

$$X^3 + PX - Q = X^3 \left\{ 1 + \frac{P}{X^2} - \frac{Q}{X^3} \right\}$$

Then when X is very large, both P/X^2 and Q/X^3 are very small so that the expression within $\{\ldots\}$ is very close to 1 and so certainly positive. Thus the sign of $X^3 + PX - Q$ is determined by that of X^3 which is > 0 for $X > 0$ and < 0 for $X < 0$. In other words, For X very large and negative, $X^3 + PX - Q$ is < 0 while for X very large and > 0, $X^3 + PX - Q$ is > 0. If we now imagine the graph of

$$Y = X^3 + PX - Q$$

we see that for X very large and < 0 it is below the x–axis while for X very large and > 0 it is above the x–axis. Thus at some point the graph must cross the x–axis, *for reasons of continuity.* This shows that there is at least one real root. We say *at least* one because it is conceivable that the graph crosses at more than one point. But it is then clear that it must cross at three points in order to end up above the x–axis. Indeed, the first crossing is from below the x–axis to above; so the second crossing, if there is one, is from above to below, and then in order to end up above the x–axis the graph has to cross the axis once again, this time from below to above, thus producing three crossings. The main point here is to show that the number of crossings is 1 or 3 according as $\Delta > 0$ or $\Delta < 0$. In the exercises below we shall see how this can be shown using a little calculus.

It is also conceivable that one of these crossings is really not a crossing but a point where the graph *touches* the x–axis; in this case the equation has two distinct real roots, but one of them is a *double, or a repeated root.* This happens when $\Delta = 0$ and the relevant calculations are indicated in the exercises below. One can also picture this in the following manner. Suppose that $\Delta = 0$. We now consider cubic equations

$$X^3 + P'X = Q'$$

where P' and Q' are close to P and Q but are such that the discriminant of this cubic equation, say Δ', is < 0, although it will be very small. We shall then find that there are three crossings, but, as P' and Q' go near P and Q, two of the crossings will tend to come together, and in the limit will coincide, producing a point of tangency with the x–axis.

Finally, when $P = Q = 0$, the graph of $Y = X^3$ crosses the x–axis at 0 and nowhere else, but 0 is a triple root.

THE POLYNOMIAL ALGEBRA

It is natural to ask whether it is possible to treat some of the above issues–such as double roots, three distinct real roots, and so on–in an *algebraic manner*, i.e., without bringing in pictures and considerations of continuity. This is possible but will require a substantial amount of work. Nevertheless we shall initiate such a treatment now, at least to give a glimpse of how the fully algebraic treatment is developed.

We consider *polynomials* in a variable which we call X. These are expressions such as
$$X^2, \quad X^3 + 3X - 1, \quad 12X^8 - (3/4)X^5 + \sqrt{8}X - \sqrt[3]{12}$$

and so on. In general we shall consider as a polynomial any expression

$$a_0 X^n + a_1 X^{n-1} + \ldots + a_{n-1} X + a_n$$

where $a_0, a_1, \ldots, a_{n-1}, a_n$ are real numbers; the a_j are called the *coefficients* of the polynomial; thus a_0 is the coefficient of X^n, a_1 is the coefficient of X^{n-1}, etc. If a term X^k is omitted, its coefficient is 0. The highest power of X which has a nonzero coefficient is said to be the *degree of the polynomial*. Thus if the degree is 2, we have a quadratic polynomial, if the degree is 3, we have a cubic polynomial, and so on.

Polynomials can be added, subtracted, and multiplied. For addition (resp. subtraction) one simply adds (resp. subtracts) the coefficients of the same powers of X, and for multiplying one multiplies every term of one of the polynomial with every term of the other polynomial and adds the coefficients of the same power of X. As an example

$$(X^2 + 21X - 18) + (X^5 - 11X + 21) = X^5 + X^2 + 10X + 3$$
$$(X^2 + 21X - 18) - (X^5 - 11X + 21) = -X^5 + X^2 + 32X - 39$$

while

$$(X^2 + 2X - 3) \times (X^5 - X + 2) = X^7 + 2X^6 - 3X^5 - X^3 + 7X - 6$$

For theoretical discussions we denote polynomials by symbols such as F, G, etc (sometimes $F(X), G(X)$, etc) and write $F + G$ and FG for their sum and product respectively. Notice that

$$F + G = G + F, F + (G + H) = (F + G) + H \overset{\text{def}}{=} F + G + H$$

$$FG = GF, F(G + H) = FG + FH, (FG)H = F(GH) \overset{\text{def}}{=} FGH$$

which resemble the properties of the numbers themselves. Finally, when we say that two polynomials F and G are *equal*, we mean that they have the same coefficients; thus $F = G$ means $F - G = 0$ where we write 0 for the polynomial all of whose coefficients are 0.

Unlike real or rational numbers, one cannot divide one polynomial by another, except in exceptional circumstances. *In this respect the polynomials resemble integers more than real numbers or rational numbers, because in general one integer does not divide another.* We shall say that a polynomial F *divides* another polynomial G if there is a third polynomial H such that

$$G = FH$$

Notice the similarity with the concept of divisibility of integers. Here we should assume, as we do in the theory of integers, that $F \neq 0$.

A polynomial is by definition an expression. As such it has "no value" but if we substitute for X a number a, we get a number. This is the process of *evaluation at* a of a polynomial. If F is the polynomial, evaluation at a gives the number which we write as $F(a)$. In particular, if $F(a) = 0$ we shall say that $X = a$ is a root of the equation $F(X) = 0$. These are of course concepts that are familiar, but we have taken the trouble to make them very precise.

Since we work with polynomials with coefficients which are real numbers, we can always divide by the coefficient of the highest power of X to make sure that this highest power coefficient is 1. Such a polynomial is called *monic*. Thus

$$X^3 - X + 18$$

monic but

$$2X^2 + 11$$

is not monic; but dividing this by 2 will yield the monic polynomial

$$X^2 + \frac{5}{2}$$

We now come to the crucial fact which relates the factorization of polynomials with roots. Consider the factorization

$$X^2 - a^2 = (X + a)(X - a)$$

with which everyone is familiar. This shows that

$$X^2 - a^2 = 0 \iff X = \pm a$$

Similarly if F is a polynomial and we have a factorization

$$F(X) = (X - a)G(X)$$

then it is immediate that

$$F(a) = 0$$

In other words, $X = a$ is a root of the equation $F(X) = 0$. Of course sometimes factorizations do not tell us what the roots are. Thus, the factorization

$$X^4 + 5X^2 + 4 = (X^2 + 1)(X^2 + 4)$$

does not tell us what the roots of the equation

$$X^4 + 5X^2 + 4 = 0$$

are, but it certainly tells us that this equation has no real root, as neither factor on the right side of the factorization vanishes for any real number.

The relationship between roots and factorizations is capable of a very simple and general formulation, called the *remainder theorem*. We shall first show that if F is a nonzero *monic* polynomial of degree ≥ 1, and t is *any real number*, we have a *factorization with remainder*:

$$F(X) = (X - t)G(X) + F(t)$$

where G is a *monic* polynomial. G is *uniquely determined*, once F and t are given. We call G the *quotient of dividing F by $X - t$*, and $F(t)$ as *the remainder*. If we assume this theorem, then it is immediate that F is divisible by $X - t$ precisely when $X = t$ is a root of $F(X) = 0$.

How does one prove the remainder theorem? The simplest method is to start with

$$F(X) = X^n + a_1 X^{n-1} + \ldots + a_{n-1}X + a_n$$

and assume that G has the form

$$G(X) = X^{n-1} + b_1 X^{n-2} + \ldots + b_{n-2}X + b_{n-1}$$

where

$$b_1, b_2, \ldots, b_{n-1}$$

are numbers to be determined. The point here is that we are looking for monic G, and as F has degree n and $X - a$ has degree 1, G is forced to have degree $n - 1$. We want the equation

$$F(X) = (X - t)G(X) + r$$

where r is a number to be determined also. We multiply and compute the right side and equate the coefficients with those of F on the left side which are all known. Equating the coefficient of X^{n-1} we get

$$a_1 = b_1 - t \implies b_1 = a_1 + t$$

which determines b_1 uniquely. The comparison of the coefficients of X^{n-2} gives

$$a_2 = b_2 - tb_1 \implies b_2 = a_2 + tb_1$$

which determines b_2 uniquely, since we know a_2 and *we have already determined* b_1. We proceed step by step and the formal argument is completed by induction. The comparison of the coefficients of $X^{n-k}(n - k \geq 1)$ gives

$$a_k = b_k - tb_{k-1} \implies b_k = a_k + tb_{k-1}$$

so that knowledge of b_{k-1} determines b_k uniquely. This process determines G completely but we are still left with the comparison of the term not involving X, the so-called *constant term*. This leads to the equation

$$a_n = r - tb_{n-1} \implies r = a_n + tb_{n-1}$$

We are done, except that the formula for r is not in the form we want. There are now two ways of showing that the expression for r is indeed the correct one. The first is to chase back the definition of $b_{n-2}, b_{n-3}, \ldots$ and show that $r = F(t)$. The other, the easier one, is by evaluation. We have

$$F(X) = (X - t)G(X) + r$$

Now evaluate both sides at $X = t$. Then, as $X - t$ becomes zero at t we have

$$F(t) = r$$

which is the assertion of the theorem.

The reader cannot fail to notice the similarity with the usual notions of quotient and remainder when we divide one integer by another. *This analogy of the polynomial algebra with the integers is one of the most basic facts of modern algebra.* One can pursue this analogy in great depth, making up definitions and checking whether expected results are indeed true. For instance, we can extend the concept of root by saying that $X = t$ is a *double root* of $F(X) = 0$ if we have a factorization

$$F(X) = (X - t)^2 H(X)$$

where H is a polynomial. Replacing $(X-t)^2$ by $(X-t)^3, (X-t)^4$, etc, we have the notion of a *triple root*, a *quadruple root*, etc. One can also introduce the notion of a *prime*! We shall say that a monic polynomial F of degree ≥ 1 is a *prime* if it is not divisible by any monic polynomial of degree ≥ 1 except for F. Traditionally such polynomials are called *irreducible*. Certainly, any polynomial $X - t$ is irreducible but so is $X^2 + 1$, as any factorization will imply one of the factors is of the form $X - t$ and this is absurd as $t^2 + 1$ cannot be 0.

At the foundation of arithmetic is the result that any positive integer can be written in a unique manner as the product of primes. For polynomials with real coefficients the corresponding theorem is that any monic polynomial (with real coefficients) can be written in a unique manner as a product of factors which are either of the form $X - t$ where t is a real number, or of the form $X^2 + 2sX + t$ where s and t are real numbers such that $s^2 < t$ (so that $X^2 + 2sX + t$ does not have a real root). This is a version of the so–called *fundamental theorem of algebra* which we shall discuss much later. It is a very deep result in algebra and was first proved by **Gauss**. We shall not go any more into the theory of polynomials at this time.

NOTES AND EXERCISES

1. Show that if $f(X) = X^3 + PX - Q$, then

$$f\left(\frac{3Q}{2P}\right) = \frac{Q}{864P^3}\left(\frac{Q^2}{4} + \frac{P^3}{27}\right) = \frac{27Q}{2P^3}\Delta$$

Hence show that if $\Delta = 0$, then $X = 3Q/2P$ is a root. Then show that we have a factorization

$$X^3 + PX - Q = \left(X - \frac{3Q}{2P}\right)^2\left(X + \frac{3Q}{P}\right)$$

by multiplying out the right side. Conclude that in this case the three roots are

$$\frac{3Q}{2P} \text{ (doubl)}, -\frac{3Q}{P}$$

2. Let f be as in the previous exercise but now assume that $\Delta \neq 0$. Show that the derivative f' of f is $3X^2 + P$.

   (i) Show that if $P > 0$, $f'(x) > 0$ for all x and hence that $f(x)$ is strictly increasing as x increases. Hence deduce that there is exactly one crossing of the x–axis by the graph of $y = f(x)$.

   (ii) From now on suppose $P < 0$, say $P = -P'$ where $P' > 0$. Show that $f'(x) = 0$ has the two solutions

   $$x_+ = +\sqrt{\frac{P'}{3}}, \qquad x_- = -\sqrt{\frac{P'}{3}}$$

   (iii) By verifying that $f'' = 6x$ show that the graph of $y = f(x)$ increases as x increases to x_-, then decreases as x increases from x_- to x_+, then increases again as x increases beyond x_+. Hence deduce that the condition for the graph of $y = f(x)$ to cross the x–axis three times is

   $$f(x_-) > 0, \qquad f(x_+) < 0$$

   (iii) Show that $f(x_-) > f(x_+)$ and hence deduce that the condition for three crossings can be written in the simplified form

   $$f(x_=)f(x_+) < 0$$

(Hint : See (iv) below for the values of $f(x_{\pm})$.)

(iv) Write $b = +\sqrt{-P/3}$ and verify that

$$f(x_{\pm}) = \mp 2b^3 - Q, \qquad f(x_-)f(x_+) = 4\Delta$$

Deduce that $\Delta < 0$ is the condition for the existence of 3 real roots of the equation $X^3 + PX = Q$; one root $< x_-$, one root between x_- and x_+, and one root $> x_+$.

3. Consider the general cubic equation

$$X^3 + AX^2 + BX + C = 0$$

(i) Write $X = Y - A/3$ and show that the equation for Y is $Y^3 + PY = Q$ where

$$P = -\frac{A^2}{3} + B, \qquad Q = -2\frac{A^3}{27} + \frac{AB}{3} - C$$

(ii) Hence show that the discriminant of the original cubic, defined as the discriminant of the cubic in Y, is

$$\frac{1}{108}\left(4B^3 - A^2B^2 + 4CA^3 + 27C^2 - 18ABC\right)$$

(iii) Deduce the criteria for the nature of the roots for the general cubic equation in terms of this discriminant.

4. Discuss the nature of the roots of the following cubic equations :

$$X^3 + X = 3, \qquad X^3 - 3X + 1 = 0, \qquad X^3 + X^2 + X + 2 = 0$$

5. Show by explicit calculation that if

$$F(X) = X^3 + AX^2 + BX + C$$

then for any real number t we have

$$F(X) = (X - t)(X^2 + RX + S) + F(t)$$

and determine R and S explicitly. Deduce that $X = t$ is a root of $F(X) = 0$ if and only if $X - t$ divides $F(X)$ exactly, namely

$$F(X) = (X - t)(X^2 + RX + S)$$

6. If $F(X)$ has three roots u, v, w, deduce from the previous exercise that

$$X^3 + AX^2 + BX + C = (X - u)(X - v)(X - w)$$

Hence obtain the formulae

$$u + v + w = -A, \qquad uv + vw + wu = B, \qquad uvw = -C$$

These are the formulae analogous to the formulae for the sum and product of the roots of a quadratic equation discussed in the text. Here again there is *Newton's theorem* : if g is any polynomial function (resp. with integer coefficients) of the roots which is *symmetric*, i.e., which is unchanged if any two of u, v, w are interchanged, then it can be expressed as a polynomial (resp. with integer coefficients) of A, B, C.

7. Obtain expressions for $u^3 + v^3 + w^3$ and $u^4 + v^4 + w^4$ in terms of A, B, C.

8. Prove that if s, t are real numbers then $X^2 + 2sX + t$ is irreducible if and only if $s^2 < t$, i.e., if and only if the equation $X^2 + 2sX + t = 0$ has no real root.

9. If F, G are polynomials of degree n, m respectively, and G is monic, prove that we can write
    $$F = H.G + R$$
    where H is a polynomial of degree $n - m$ ($H = 0$ if $n < m$) and R is a polynomial of degree $< m$. Show also that H and R are uniquely determined by F and G.

10. Show that the polynomial $X^3 + X + 1$ cannot be written as a product of polynomials *with rational coefficients* of lower degree.

ARS MAGNA AND ITS ROLE IN THE

BIRTH OF MODERN ALGEBRA

We shall now look more closely into Cardano's great treatise *Ars Magna* so as to enable us to understand the significant role it played in the development of Modern Algebra and also appreciate why Cardano should be regarded as a central figure in the birth of modern algebra.

First of all there is the matter of proper acknowledgement. Cardano does not hide anything and describes with complete honesty how he came into possession of the solution for the equation
$$X^3 + PX = Q$$
and how he was led to his own and Ferrari's results as a consequence of their explorations. Let us listen to his own words ([RW], p. 8) :

In our own days Scipione del Ferro of Bologna has solved the case of the cube and first power equal to a constant, a very elegant and admirable accomplishment. Since this art surpasses all human subtlety and the perspicuity of mortal talent and is a truly celestial gift and a very clear test of the capacity of men's minds, whoever applies himself to it will believe that there is nothing that he cannot understand. In emulation of him, my friend Niccolò Tartaglia of Brescia, wanting not to be outdone, solved the same case when he got into a contest with his [Scipione's] pupil Antonio Maria Fior, and moved by my many entreaties, gave it to me. For I had been deceived by the words of Luca Pacioli, who denied that any more general rule could be discovered than his own. Notwithstanding the many things which I had already discovered, as is well known, I had despaired and had not attempted to look any further. Then, however, having received Tartaglia's solution, and seeking for the proof of it, I came to understand that that there were a great many other things that could also be had. Pursuing this thought with increased confidence, I discovered these others, partly by myself and partly through Lodovico Ferrari, formerly my pupil. Hereinafter those things which have been discovered by others have their names attached to them; those to which no name is attached are mine. The demonstrations, except for the three by Mahomet and the two by Lodovico, are all mine. Each is individually set out under a proper heading and, following the rule, an illustration is added.

Later on when he discusses the cubic in the form assumed by Scipione del Ferro, he has this to say ([RW], p. 96): *Scipione del Ferro of Bologna well-nigh thirty years ago discovered this rule and handed it on to Antonio Maria Fior of Venice, whose contest with Noccolò Tartaglia gave Niccolò occasion to discover it. He [Tartaglia] gave it to me in response to my entreaties, though withholding the demonstration. Armed with this assistance, I sought out the demonstration in [various] forms.*

If all Cardano had done was to give an exposition of the discoveries of del Ferro and Tartaglia, he would not be regarded with the reverence he is given today; nor would that book be regarded as signaling the birth of modern algebra. What Cardano did was far more remarkable. His curious and encyclopedic mind realized that there were other forms of the cubic, including those which had the X^2 term present, and that it was an open question as to how their solutions should be worked out. He realized, much more clearly than anyone before him, that in these other cases the cubic will have, not one, but three, solutions, and so a single formula cannot account for all of these without further analysis. He was the first one to have a global view of the totality of all cubic equations and realized how one can go from one equation to another by simple transformations; he was especially fond of the symmetry

$$X \longrightarrow -X$$

that would change positive (real) solutions to negative (false or fictitious) ones and vice versa. He knew the relation between the roots and the coefficients such as the result that for the cubic

$$X^3 + AX^2 + BX + C = 0$$

the sum of the roots is

$$-A$$

Above all, he realised that the formula of del Ferro and Tartaglia led to square roots of negative numbers, knew when this would happen (in modern terminology, when the discriminant is negative), and discovered the paradoxical situation (the so-called *irreducible case*) where all the roots are real but the formula involves square roots of negative numbers. He discovered that if one manipulates with these quantities *putting aside the mental tortures involved* ([RW], pp. 219—220), one can verify the fact that these expressions describe the roots, a feat that he called *truly sophisticated*. In this way he must be regarded as one of the first to encounter the need to introduce complex numbers, something that was not done till centuries later, and had to wait till Gauss.

It is abundantly clear even from this brief description that *Ars Magna* introduced a dramatic change in the level of discourse and study in algebra from its predecessors. It pointed the way to the formulation and solution of *general* questions, showed that the theory of the cubic and biquadratic equations was essentially complete, and that one had to introduce extensions of the number systems to reach a real understanding and make further progress. The fact that he manipulated with symbols involving quantities not properly defined, such as square roots of negative numbers, marks him out as one of the first truly original algebraists; it was only in this century, and that too under the impact of discoveries in modern physics, especially quantum theory, that it became clear that *the role of Algebra was to explore and create all possible number systems so that among them can be found those that mirror what happens in the real world*. This point of view is most beautifully explored in the recent book by one of the great twentieth century masters of Algebra, **I. R. Shafaraevich** (see *Algebra, Vol 11, Encyclopedia of Mathematical Sciences, Springer Verlag, 1990*).

9

SOLUTION OF THE BIQUADRATIC EQUATION

FERRARI'S METHOD

We now turn to the equation of degree 4 which in general has the form

$$X^4 + AX^3 + BX^2 + CX + D = 0$$

However we have already seen in the exercises to the preceding chapter that sometimes these equations do not have any real root. Thus, for the equation

$$X^4 + 4X^2 + 3 = 0$$

there is no real root. To see this we observe first that we have a factorization

$$X^4 + 4X^2 + 3 = (X^2 + 3)(X^2 + 1)$$

of the fourth degree polynomial as a product of two quadratic polynomials. Certainly, $X = t$ is a root of the equation above if and only if either $t^2 + 3 = 0$ or $t^2 + 1 = 0$, which is not possible for real t. So, what is the sense in which Ferrari solved the fourth degree equation? The answer, as we shall see, is that his method depends on solving an auxiliary cubic equation first, and after finding a root of this cubic, to solve two quadratic equations. In the first step it is perfectly possible to encounter the irreducible case, so that the root may not be expressible in real terms; even if the first step leads to a case when the discriminant of the cubic is ≥ 0, the second step may involve negative numbers inside the root sign. Thus it is clear that any interpretation of Ferrari's solution has to rely on complex numbers. Of course it is possible that the method encounters neither obstacle–in other words, the first step leads to a cubic with positive discriminant, and the second step leads to quadratics which again have positive discriminants. We shall see that there are examples of all these possibilities. In spite of these qualifications, Ferrari's achievement was a great one and put the finishing touches to the great accomplishment of Italian algebraists that led to the creation of modern algebra.

REMOVAL OF THE X^3 TERM

The first step is exactly as in the theory of the cubic equation, namely to pass to another equation that has no term involving X^3. We do it in the same way. We write

$$X = Y + t$$

so that the equation in Y has no Y^3 term for an appropriate choice of t. Now

$$X^4 + AX^3 + BX^2 + CX + D = (Y + t)^4 + A(Y + t)^3 + B(Y + t)^2 + C(Y + t) + D$$

and we calculate what the coefficient of Y^3 is on the right side. Since

$$(Y + t)^4 = (Y + t)(Y + t)^3 = (Y + t)(Y^3 + 3Y^2t + \ldots), \quad A(Y + t)^3 = AY^3 + \ldots$$

it follows that the coefficient of Y^3 in the right side is

$$4t + A$$

which will be zero if we choose

$$t = -\frac{A}{4}$$

In practice we have to complete the calculation of all the coefficients in the polynomial in Y to get the full expression in terms of Y. Moreover, $X = x$ is a root of the equation in X if and only if $Y = x - t$ is a root of the equation in Y. In other words, the roots of the equation in X are obtained by adding t to the roots of the equation in Y.

As an example, let us take the equation

$$X^4 - 2X^3 - X^2 + 2X = 0$$

whose roots are $X = -1, 0, 1, 2$ as may be easily verified. In fact we have the factorization

$$X^4 - 2X^3 - X^2 + 2X = X(X + 1)(X - 1)(X - 2)$$

The substitution $X = Y + 1/2$ leads to the equation

$$(Y + \frac{1}{2})(Y + \frac{3}{2})(Y - \frac{1}{2})(Y - \frac{3}{2}) = Y^4 - \frac{5}{2}Y^2 + \frac{9}{16} = 0$$

THE AUXILIARY CUBIC EQUATION

We shall now assume that the equation to be solved is of the form

$$X^4 + AX^2 + BX + C = 0$$

The idea behind Ferrari's method is to introduce a variable u and rewrite the above equation by completing squares. Since

$$X^4 + AX^2 = \left(X^2 + \frac{1}{2}A + u\right)^2 - 2uX^2 - \left(\frac{A}{2} + u\right)^2$$

the equation for X becomes

$$\left(X^2 + \frac{A}{2} + u\right)^2 = 2uX^2 - BX + \left(u^2 + Au - C + \frac{A^2}{4}\right)$$

The variable u is at our disposal. *We now choose it so that the right side is a perfect square.* For this to be true, the discriminant of the quadratic expression in X must be zero. Now, the discriminant of

$$RX^2 + SX + T$$

is

$$S^2 - 4RT;$$

and if $S^2 - 4RT = 0$, we can express $RX^2 + SX + T$ as

$$RX^2 + SX + T = R\left(X + \frac{S}{2R}\right)^2$$

We must therefore choose u so that

$$B^2 - 4(2u)(u^2 + Au - C + (A^2/4)) = 0$$

In other words, we must choose u to be a solution of the *cubic* equation

$$8u^3 + 8Au^2 + (2A^2 - 8C)u - B^2 = 0$$

This is the *auxiliary cubic equation.* In principle it can be solved, by first removing the u^2 term, and then using Cardano's formula.

FERRARI'S SOLUTION

Let u_0 be a root of the auxiliary cubic equation. Then the quadratic expression in X becomes

$$2u_0\left(X - \frac{B}{4u_0}\right)^2$$

Thus, the equation for X becomes

$$\left(X^2 + \frac{A}{2} + u_0\right)^2 = 2u_0\left(X - \frac{B}{4u_0}\right)^2$$

This is equivalent to the two quadratic equations

$$\left(X^2 + \frac{A}{2} + u_0\right) = \pm\sqrt{2u_0}\left(X - \frac{B}{4u_0}\right)$$

Each of these equations, when solved, will lead to two roots and so we have a method of finding the four roots of the original biquadratic equation.

Clearly, Ferrari's method leads to a very complicated formula–we can obtain it by writing Cardano's expression for u_0 and then writing down the solutions of the two quadratic equations. But this is manifestly not worth the trouble. In any given case, the method can be applied following the steps outlined in an unambiguous manner. On the other hand, theoretically, it is a great accomplishment because it showed that there is in principle a technique for solving any biquadratic equation. Ferrari's solution is thus a climax of the development, started by Scipione del Ferro, and continued by Tartaglia and Cardano.

Let us consider the example of the equation

$$X^4 + 2X^2 - 2X - 1 = 0$$

for which $X = 1$ is an obvious root but the other roots are not rational. Indeed, we have the factorization

$$X^4 + 2X^2 - 2X - 1 = (X - 1)(X^3 + X^2 + 3X + 1)$$

and one can show that no rational number can be a root of the equation

$$X^3 + X^2 + 3X + 1 = 0$$

(See exercise below). The original equation already has no X^3 term and so we can directly proceed to the formation of the auxiliary cubic equation. Since

$$X^4 + 2X^2 = (X^2 + (1 + u))^2 - 2uX^2 - (1 + u)^2$$

the equation to be solved becomes

$$(X^2 + (1 + u))^2 = 2uX^2 + 2X + (1 + u)^2 + 1$$

We now choose u so that the quadratic expression on the right side is a square, and the condition for this becomes

$$2u^3 + 4u^2 + 4u - 1 = 0$$

or

$$u^3 + 2u^2 + 2u - \frac{1}{2} = 0$$

To solve this we must remove the u^2 term. Write

$$u = v - \frac{2}{3}$$

Then the equation for v becomes

$$v^3 + \frac{2}{3}v - \frac{67}{54} = 0$$

which is of the type solved by del Ferro. There is exactly one real root, say v_0 and so exactly one real root, say u_0 for the equation in u. Notice that this root u_0 is > 0 and $< 1/4$ because the expression

$$2u^3 + 4u^2 + 4u - 1$$

is equal to -1 when $u = 0$ and equal to $9/32$ when $u = 1/4$ so that it must vanish between 0 and $1/4$. For this choice of $u = u_0$, the equation in X becomes

$$(X^2 + (1 + u_0))^2 = 2u_0 \left(X + \frac{1}{2u_0} \right)^2$$

which leads to the two quadratic equations

$$X^2 - \sqrt{2u_0}X + 1 + u_0 - \frac{1}{\sqrt{2u_0}} = 0, \qquad X^2 + \sqrt{2u_0}X + 1 + u_0 + \frac{1}{\sqrt{2u_0}} = 0$$

If we substitute $X = 1$ in the first of these we get $2u_0^3 + 4u_0^2 + 4u_0 - 1 = 0$ as we should so that $X = 1$ is indeed a root. Since 1 is a root of the first quadratic and the sum of its roots is $\sqrt{2u_0}$ the four roots are

$$1, \qquad \sqrt{2u_0} - 1, \qquad \frac{1}{2}\left(-\sqrt{2u_0} \pm \sqrt{-2u_0 - 4 - \frac{4}{\sqrt{2u_0}}} \right)$$

Using del Ferro's formula for v_0 we have

$$u_0 = v_0 - \frac{2}{3}, \qquad v_0 = \sqrt[3]{\sqrt{\frac{19}{48}} + \frac{67}{108}} - \sqrt[3]{\sqrt{\frac{19}{48}} - \frac{67}{108}}$$

It will be clear from the exercises below that all possibilities can occur while using this method.

NOTES AND EXERCISES

1. Prove that $X^3 + X^2 + 3X + 1 = 0$ has no rational roots.

2. Consider the equation
$$X^4 + 2aX^2 + 4bX + a^2 = 0$$

where a, b are real numbers. Show that the auxiliary cubic equation is

$$u^3 + 2au^2 - 2b^2 = 0$$

Show that the substitution $u = v - (2a)/3$ converts this into the equation

$$v^3 - \frac{4a^2}{3}v = 2b^2 - \frac{16a^3}{27}$$

Show that the discriminant of this cubic equation is

$$\Delta = b^2 \left(b^2 - \frac{16a^3}{27} \right)$$

Deduce that for the auxiliary cubic there is a unique real root (resp. three real roots) if and only if

$$27b^2 - 16a^3 > 0 \ (\text{ resp. } < 0)$$

Hence give examples of both of these possibilities by giving special values for a, b.

3. Write

$$X^4 + AX^3 + BX^2 + CX + D = (X - t_1)(X - t_2)(X - t_3)(X - t_4)$$

and deduce the following relations between symmetric functions of the t_i's and the coefficients:

$$t_1 + t_2 + t_3 + t_4 = -A$$
$$t_1t_2 + t_2t_3 + t_3t_1 + t_1t_4 + t_2t_4 + t_3t_4 = B$$
$$t_2t_3t_4 + t_1t_3t_4 + t_1t_2t_4 + t_1t_2t_3 = -C$$
$$t_1t_2t_3t_4 = D$$

From these formulae deduce expressions for

$$t_1^2 + t_2^2 + t_3^2 + t_4^2, \text{ and } t_1^4 + t_2^4 + t_3^4 + t_4^4$$

in terms of A, B, C, D.

SOME THEMES FROM MODERN ALGEBRA

10

NUMBERS, ALGEBRA, AND THE REAL WORLD

There is a remark attributed to the great 19th century mathematician **Kronecker** (1823–1891) that *God made the natural numbers and all the rest were made by man.* This seemingly irreverent remark contains an enormous truth. Indeed, when properly examined, it highlights the basic fact that mathematical objects are the result of *free creation*. It is often the case that the process of creation is a response to some profound needs of the real world. But this is not always true; there are remarkable instances where, concepts created and studied originally for their beauty and simplicity were later found to be the precise tools for understanding the physical world. Two of the most celebrated examples of this are *Riemannian geometry*, which was created by **Riemann** (1826–1866) and became the indispensable tool in physics when **Einstein** discovered *Relativity theory*, and *infinite dimensional linear algebra*, whose foundations were laid by **Fredholm, Riesz, Frechet, Hilbert** and many others long before it became the essential tool in the theory of atomic systems and other aspects of quantum physics.

It may be said with some justification that discoveries about numbers and their properties were originally motivated to a great extent by the need to measure things around us and to help us in activities such as commerce and navigation. Sometime in the distant past people found that it was essential to create the concept of zero and negative numbers to bring order and economy in the description of the external world, and so the number system was extended to include the negative integers first, and then, fractions. The Pythagoreans then discovered irrational numbers as a result of geometric constructions, and it gradually became clear that one had to extend the number system widely in order to comprehend the various phenomena around us. The real number system as a far-reaching enlargement of the system of rational fractions emerged in stages spread over many centuries. It was only in the later part of the 19th century that mathematicians were able to give a proper foundation to the concept of real numbers; this came about as a consequence of the work of **Dedekind** (1831–1916). Nevertheless the real number system as a working tool was already available at least in some primitive form going back to ancient times: as examples one can cite the work of **Archimedes** on the approximations to π and the work of **Fibonacci** on the approximations to roots of certain cubic equations, as well as ancient algorithms for extracting square and cube roots.

The development of the theory of equations introduced a new kind of problem involving the concept of numbers. For instance, the formula for the solution of a quadratic equation involved taking a square root, and it became clear that in principle one can have a negative number appear inside the square root sign. Of course in any problem where the unknown X had a physical interpretation, this did not happen, so that the problem was not examined seriously till Cardano came across it in his treatment of the cubic equation. Here he discovered what is called nowadays *the irreducible case*, in which the cubic equation has three *real* solutions, but the formula involves square roots of negative numbers. This led to two very vexing questions: how to interpret these quantities, and how to make sure that the formula describes all three solutions. It took mathematicians a very long time to understand the issues involved in extending the real number system to include these new types of numbers, the so-called *complex numbers*. It was only after the work of the great mathematician **Gauss** (1777–1855) that it became possible to work comfortably with complex numbers and recognize the remarkable fact that in a certain sense the complex number system was the end of an evolutionary line of development.

More remarkably, the story of algebra took a dramatic turn in the 19th century when it was realised that other algebraic systems of great variety and subtlety could be developed and that these new systems were potentially as useful as the system of real and complex numbers. Indeed the 19th century became the golden age of algebra, and things became even better in the current century when such fantastic physical theories as quantum mechanics were found to depend on the esoteric algebraic systems that mathematicians had created while toiling in relative obscurity decades earlier.

Nowadays these issues are very well understood. The different roles of the mathematician and the physical scientist in the understanding of the real world are quite clearly perceived. The mathematician creates models and systems, sometimes because there is a need for them, and at other times because of the beauty and simplicity and inevitability of the mathematics involved. As the process of discovery leads us to the problems of understanding aspects of life that are increasingly remote from *direct* physical experience, it has been found that it is only in the world of mathematics that one can find structures with the necessary abstraction and versatility that allow one to model the physical world in all its complexity, from the innermost recesses of the atom to the outermost reaches of the universe.

11

COMPLEX NUMBERS

THE REAL NUMBER SYSTEM

We shall now begin discussing the question of extending the system of real numbers so that we can work with such strange objects as square roots of negative numbers. Of course this requires that we have a good understanding of what real numbers are. I have already mentioned that this is a difficult problem. But we do not really need such a precise understanding of the real number system. For getting an insight into how the extension to include imaginary quantities is done, it is enough to have only a rough idea of the properties of the real number system.

Let us review what we have seen about real numbers. Certainly we have the set of all rational numbers, which is denoted by $\mathbf{Q}$. On $\mathbf{Q}$ we have the four operations : addition, subtraction, multiplication, and division by non zero numbers. These operations satisfy well-known properties which everyone learns very early in one's education. But one cannot stop with the rational numbers–we have seen that geometrical constructions force us to consider irrational numbers like $\sqrt{2}, \sqrt{1+\sqrt{3}}$ and so on. Once one introduces $\sqrt{2}$ for instance, one has to consider all numbers of the form $a + b\sqrt{2}$ where a and b are rational numbers, because these will arise as soon as we start performing algebraic operations involving rational numbers and $\sqrt{2}$. But we cannot stop here; we have to introduce numbers which are solutions of polynomial equations whose coefficients are numbers we have already introduced. Even this is not enough; we have seen that numbers like $\pi, e, 2^{\sqrt{2}}$ and so on, are not roots of any equation with rational coefficients, or even algebraic coefficients. We must include in our real number system all such transcendental numbers.

Where does one stop? The solution to this problem was given by **Dedekind**. He visualized the real number system as a set of objects, denoted by $\mathbf{R}$, and having the following properties:

1. **R** contains **Q**, i.e., all rational numbers are to be included in **R**.

2. There is a notion of ordering in **R**, which allows us to decide, given any two elements a, b of **R**, whether $a < b$, or $b < a$ (which is the same as $a > b$), or $a = b$; it is also assumed that exactly one of these three possibilities is always true.

3. The notion of order is compatible with the algebraic operations. This means roughly that if $a < b$, then $a + c < b + c$ for any c (a, b, c are all in **R**), $a > 0$ is equivalent to $-a < 0$, and $a > 0, b > 0$ implies $ab > 0$ and so on.

4. The system **R** is *complete* in a technical sense. This means roughly that if we are given any set A of real numbers such that for some real number a one has $x < a$ for all x in A, there is a number b such that every number of A is $\leq b$ and no number $< b$ has this property.

We have not tried to make this list of properties absolutely precise because that is not necessary for our purposes. But first of all note that the rational numbers have the properties 1.–3. but not 4. To see this consider the number $\sqrt{2}$. We know this is not in **Q**. But if we take A to be the set of all rational numbers x such that $x > 0$ and $x^2 < 2$, then A has the property required of it in 4. For instance, all elements of A are < 2. But there is no rational number with the property demanded of b. This is because if c is any rational number which is $> x$ whenever x is rational and $x^2 < 2$, then c^2 must be > 2 and one can certainly find a rational number d such that $d < c$ and $d^2 > 2$. Thus unless $\sqrt{2}$ is added to our number system, A cannot satisfy the property 4. This is only one example. The real impact of property 4. is that it *anticipates basically all possible situations in which additional numbers have to be created, and supplies them.*

It is natural to wonder if the process of starting from **Q** and adding additional numbers will in the end produce *two distinct number systems that both possess the properties listed above.* But this cannot happen; there is essentially only one way to enlarge the rational number system so that the enlarged system has the structures of addition, subtraction, multiplication, and division by nonzero elements, and the structure of an order, and is complete in the above sense. This unique system is called *the real number system.*

This discussion is very abstract and perhaps difficult to assimilate when one encounters it for the first time. But a strict understanding of this is not needed for going on and introducing the complex number system. This is what we propose to do now.

NOTES AND EXERCISES

1. Show that if we write $\mathbf{Q}(\sqrt{2})$ for the set of real numbers of the form $a + b\sqrt{2}$ with $a, b \in \mathbf{Q}$, then it has the following properties:

    (a) If $a + b\sqrt{2} = 0$ where $a, b \in \mathbf{Q}$, then $a = b = 0$ and that this is equivalent to $a^2 - 2b^2 = 0$.

    (b) Show that if $x = a + b\sqrt{2} \neq 0$, then $1/x$ is also in $\mathbf{Q}(\sqrt{2})$ and is given by

    $$\frac{1}{x} = \frac{a}{a^2 - 2b^2} - \frac{b}{a^2 - 2b^2}\sqrt{2}.$$

2. Let A be the set of all positive rational numbers x such that $x^2 < 2$. There are numbers in A; for instance $1 \in A$. Show that if x is in A, there is a rational number y such that $y > x$ and y is still in A. Similarly, show that if z is a rational number which is positive and $z^2 > 2$, there is a rational number t such that $0 < t < z$ and $t^2 > 2$. Deduce that there is no rational number b with the property 4 with respect to the set A.

THE ALGEBRA OF COMPLEX NUMBERS

The real numbers have the following property:

$$a_1, a_2, \ldots, a_k \geq 0, a_1 + a_2 + \ldots + a_k = 0 \implies a_1 = a_2 = \ldots = a_k = 0$$

In other words, a sum of positive numbers can never be zero. Since squares are positive numbers unless they are zero, we can reformulate this as

$$b_1^2 + b_2^2 + \ldots + b_k^2 = 0 \implies b_1 = b_2 = \ldots = b_k = 0$$

In particular, we cannot solve the equation

$$X^2 + 1 = 0$$

for a real number X. More generally for any real number $a > 0$ we cannot solve

$$X^2 + a = 0$$

as a real number X, i.e., $\sqrt{-a}$ does not make sense inside $\mathbf{R}$. The real number system has to be enlarged once again if we are to make sense of these symbols.

Remarkably, it turns out that in order to work with square roots of all negative real numbers it is enough to introduce just one of them, the square root of -1; actually there should be two square roots, and we shall choose one of them and call it i; the other will be written as $-i$, and after the number system is introduced, it will be interpreted as the negative of i. We write $\mathbf{R}$ for the system of all real numbers, and *define* a complex number as an expression of the form

$$z = a + ib, \qquad (a, b \in \mathbf{R})$$

The number a is called the *real part* of the complex number z, and the number b is called the *imaginary part* of the complex number z (although it is every bit as real as a; this is just traditional terminology and there is no good reason to depart from it). If we say $z = a + ib$ is a complex number, it is implicitly assumed that a and b are real. We write $\mathbf{C}$ for the system of all complex numbers. Two complex numbers

$$z = a + ib, \quad z' = a' + ib'$$

are said to be equal,

$$z = z'$$

in symbols, if and only if $a = a', b = b'$:

$$a + ib = a' + ib' \iff a = a', b = b'$$

All arithmetical conventions retain their validity and so we shall identify the real number a with the complex number $a + i0$; thus 0 is the same as $0 + i0$, and i is the same as $0 + i1$.

The next step is to define the arithmetical operations that can be performed on the complex numbers. This means we have to define how two complex numbers are to be added and subtracted, how they can be multiplied, and how they can be divided. The rules of addition, subtraction, and multiplication are the ones given below:

$$(a + ib) \pm (a' + ib') = (a \pm a') + i(b \pm b')$$
$$(a + ib) \times (a' + ib') = (aa' - bb') + i(ab' + ba')$$

Although the rule for multiplication looks strange, it is the one we should use if we multiply the two expressions $a + ib$ and $a' + ib'$ with the convention that whenever we get i^2 we replace it by -1. Also, as is customary when dealing with multiplication of real numbers, we shall omit the multiplication sign; thus zz' is the product $z \times z'$ of the complex numbers z and z'. Some natural properties which are easy to check from the definitions are

$$z1 = z, \qquad z0 = 0, \qquad i^2 = -1, \qquad (-i)^2 = -1$$

From our definitions we know that a complex number

$$z = a + ib$$

is different from 0 if and only if at least one of a and b is different from zero. An elegant way to reformulate this is to say that

$$a + ib = 0 \iff a^2 + b^2 = 0 \iff a = b = 0$$

It is now a trivial matter to check that the rules for addition and multiplication of complex numbers *extend* the rules we already know for real numbers. But what is less trivial is that *in the system of complex numbers division by a nonzero element is always possible, exactly as it is the case for real numbers.* In fact, let us take a complex number $z = a + ib$ different from 0, so that $a^2 + b^2$ is different from 0. It is then required to find the complex number z' which, when multiplied by z, gives 1. We shall find that there is only one such complex number, so that it is entitled to be called the *inverse* of z and denoted by

$$\frac{1}{z} \text{ or } z^{-1}$$

If we take

$$z' = a' + ib'$$

and write out the condition that

$$zz' = 1 \iff (aa' - bb') + i(ab' + ba') = 1$$

we get the equations

$$aa' - bb' = 1, \qquad ba' + ab' = 0$$

which can be solved for a' and b'. We do this and obtain

$$a' = \frac{a}{a^2 + b^2}, \qquad b' = -\frac{b}{a^2 + b^2}$$

If we remember that z was different from 0 so that $a^2 + b^2$ is different from 0, it is clear that these formulae make sense and show at the same time that there is one and only one z' satisfying the requirement that $zz' = 1$. Thus

$$\frac{1}{a + ib} = \frac{a}{a^2 + b^2} - \frac{b}{a^2 + b^2} i \qquad (z \neq 0 \iff a^2 + b^2 \neq 0)$$

With this we have completed the definition of the system of complex numbers and the arithmetical operations that can be performed with them. As we have mentioned earlier, the first person who was truly comfortable working with complex numbers and explored them systematically was **Gauss** (1777-1855). One has to be cautious in working with these new numbers; for instance it is possible to have an equation like

$$z^2 + z'^2 = 0$$

without z or z' being 0, as in

$$i^2 + 1 = 0$$

If we had been very conservative and used only *rational fractions* as the real and imaginary parts of the complex numbers, everything above will make sense, and the system we get may be called the system of *Gaussian rationals*, in honour of Gauss, and denoted by $\mathbf{Q}(i)$; if the real and imaginary parts are restricted to be *integers*, the system we get is called the system of *Gaussian Integers*. As in the case of ordinary integers, division is not always possible in the system of Gaussian integers.

CALCULATIONS WITH COMPLEX NUMBERS

We shall do now some examples of calculations with complex numbers.

addition

$$1 + 2i + 2 - 3i = 3 - i$$

$$3 - 6i - (4 + 5i) = -1 - 11i$$

multiplication

$$7 - 6i \times 3 + 2i = (21 + 12) + i(-18 + 14) = 33 - 4i$$

$$(-i)^3 = -i^3 = -(-1)i = i$$

$$\left(\frac{\sqrt{3} + i}{2}\right)^3 = \frac{1}{8}(\sqrt{3} + i)^3$$

$$= \frac{1}{8}(\sqrt{3} + i)^2(\sqrt{3} + i)$$

$$= \frac{1}{8}(2 + 2i\sqrt{3})(\sqrt{3} + i)$$

$$= \frac{1}{8}(2\sqrt{3} - 2\sqrt{3} + 6i + 2i)$$

$$= i$$

$$\left(\frac{-\sqrt{3} + i}{2}\right)^3 = \frac{1}{8}(2 - 2\sqrt{3}i)(-\sqrt{3} + i)$$

$$= i$$

These calculations show that we have found three *complex cube roots* of i:

$$\sqrt[3]{i} = -i, \quad \left(\frac{\sqrt{3}+i}{2}\right), \quad \left(\frac{-\sqrt{3}+i}{2}\right)$$

Actually there are no others because

$$X^3 - i = (X+i)\left(X - \frac{\sqrt{3}+i}{2}\right)\left(X - \frac{-\sqrt{3}+i}{2}\right)$$

which is easily verified.

Solution of $X^3 - X = 0$

The roots are

$$X = 0, \quad 1, \quad -1$$

since

$$X^3 - X = X(X^2 - 1) = X(X+1)(X-1)$$

Let us now show that we can use Cardano's formula to reach this result. *This is important; because it will show that even when the formula involves square roots of negative numbers, one can calculate with complex numbers to find the actual solutions which are real.* The formula gives

$$\Delta = -\frac{1}{27}, \qquad \sqrt{\Delta} = \frac{1}{3\sqrt{3}}i$$

so that

$$X = \sqrt[3]{\frac{i}{3\sqrt{3}}} - \sqrt[3]{\frac{i}{3\sqrt{3}}}$$

Since

$$\sqrt[3]{3\sqrt{3}} = \sqrt{3}$$

this can be written as

$$X = \frac{1}{\sqrt{3}}\left(\sqrt[3]{i} - \sqrt[3]{i}\right)$$

The great temptation is to say at this stage that X has to be 0. But we should not be hasty; the cube root sign does not have a single answer, but three! So this expression will give altogether $3 \times 3 = 27$ values for the expression for X! But a calculation shows that only 3 of them will be real. These are

$$0, \quad 1, \quad -1$$

obtained when we take for the two cube roots the three pairs of values

$$-i, -i; \quad \frac{\sqrt{3}+i}{2}, \frac{-\sqrt{3}+i}{2} ; \quad \frac{-\sqrt{3}+i}{2}, \frac{\sqrt{3}+i}{2}$$

We shall discuss later on the general cubic equation and prove that in all cases one has an interpretation in terms of complex numbers which shows that the formula of Cardano will give us the roots in every case. Thus the insights and speculations of Cardano are finally justified.

Fourth roots

Here is an example where fourth roots are computed:

$$\sqrt[4]{-1} = \frac{1+i}{\sqrt{2}} , \quad \frac{-1+i}{\sqrt{2}} , \quad -\frac{1+i}{\sqrt{2}} , \quad \frac{1-i}{\sqrt{2}}$$

We shall see later on that the well-known trigonometric identities for

$$\cos(A+B), \quad \sin(A+B)$$

will allow us to determine the Nth roots of complex numbers.

GEOMETRY OF COMPLEX NUMBERS

REAL LINE AND COMPLEX PLANE

It is understandable if there is a slightly dazed feeling at this stage. The one act of writing a symbol i seems to have created a new universe of calculations, provided we follow the rules for calculating with the newly created numbers. Let me point out that this type of definition should not really be new to anyone; indeed, it is not very different from the way fractions are defined. A positive fraction is nothing more than a symbol

$$\frac{a}{b} , a, b \text{ natural numbers}$$

where the rules of calculation with these symbols are as follows:

$$\frac{a}{b} = \frac{ka'}{kb'}$$
$$\frac{a}{b} + \frac{a'}{b'} = \frac{ab' + a'b}{bb'}$$
$$\frac{a}{b} \times \frac{a'}{b'} = \frac{aa'}{bb'}$$

What we have done for complex numbers is no different. Indeed, we can even dispense with the symbol i by writing

$$(a, b)$$

instead of

$$a + ib$$

and define the rules of addition, subtraction, and multiplication by

$$(a, b) \pm (a', b') = ((a \pm a'), (b \pm b'))$$
$$(a, b) \times (a', b') = ((aa' - bb'), (ab' + ba'))$$

We then observe that the numbers of the form

$$(a, 0)$$

have the same rules as the real numbers

$$a$$

and so we can *identify* a and $(a, 0)$. Moreover the rule for multiplication gives

$$(0, 1)^2 = (-1, 0)$$

So if we now introduce the *notation*

$$i = (0, 1)$$

and remember that we have agreed to treat $(a, 0)$ and a as the same, we get the equation

$$i^2 = -1$$

At the same time the relation

$$(a, b) = (a, 0)(1, 0) + (0, 1)((b, 0)$$

which follows from the rules shows that we can write

$$(a, b) = a + ib$$

which was the original definition!

In high school one has learned to represent points in the plane by *ordered pairs* of real numbers. Thus

$$(a, b)$$

represents the point whose X-coordinate is a and Y-coordinate is b. *We can obviously use the same convention for representing complex numbers.* Thus the complex numbers will he represented by points of the plane; the point

$$(a, b)$$

represents the complex number

$$a + ib$$

The plane with this interpretation is called the *complex plane*. The points of the X-axis then represent the *real numbers* while the points of the Y-axis represent numbers of the form

$$ib \quad (b \text{ real })$$

which are called *pure imaginary*. The X-axis, when viewed by itself with the interpretation that its points represent real numbers, is called the *real line*.

Given a complex number

$$z = a + ib \quad (a, b \in \mathbf{R})$$

we define its *conjugate* as the complex number

$$\bar{z} = a - ib$$

The process of passing from z to its conjugate $\bar{z}$ is called *complex conjugation*. Geometrically, this process is just reflection in the real axis. Let us now see what happens if we multiply z and $\bar{z}$. We get

$$z\bar{z} = a^2 + b^2 \qquad (z = a + ib, a, b \in \mathbf{R})$$

So if we define

$$|z| = +\sqrt{a^2 + b^2}$$

then $|z|$ is the *distance* of the point z from the origin $0 = (0,0)$. $|z|$ is called the *absolute value* of z. The following properties of the absolute value and complex conjugation are easy to derive.

$$|zz'| = |z||z'|$$
$$|z^{-1}| = |z|^{-1} \quad (z \neq 0)$$
$$\left|\frac{z}{z'}\right| = \frac{|z|}{|z'|} \quad (z' \neq 0)$$
$$\overline{zz'} = \bar{z}\bar{z'}$$

Finally, the real part of z is denoted by $\Re(z)$ and the imaginary part of z by $\Im(z)$. We have

$$\Re(z) = \frac{1}{2}(z + \bar{z})$$

$$\Im(z) = \frac{1}{2i}(z - \bar{z})$$

NOTES AND EXERCISES

1. Check the following for elements z, z', z'', t, u in $\mathbf{C}$:

$$z + z' = z' + z, \quad (z + z') + z'' = z + (z' + z'')$$

$$zz' = z'z, \quad z(z'z'') = (zz')z'', \quad z(t + u) = zt + zu$$

2. Prove the identities about complex conjugation and absolute value stated in the text. Prove also that $|z - z'|$ is the distance between the points z and z' of $\mathbf{C}$.

3. For $z \in \mathbf{C}$ let $N(z) = z\bar{z}$. Show that $N(z)$ is a real number which is always ≥ 0, and that $N(z) = 0$ if and only if $z = 0$. Show also that $N(zz') = N(z)N(z')$.

4. Let $\mathbf{Z}[i]$ denote the set of Gaussian integers, i.e., the set of all complex numbers of the form $z = a + ib$ where $a, b \in \mathbf{Z}$, $\mathbf{Z}$ being the set of integers. If $z_1 \neq 0, z_2 \in \mathbf{Z}[i]$ we say that z_1 *divides* z_2 if there is a z_3 in $\mathbf{Z}[i]$ such that $z_2 = z_1 z_3$.

   (a) For $z \in \mathbf{Z}[i]$ show that $N(z)$ is an integer ≥ 0 and that if $z_1(\neq 0)$ divides z_2 $(z_i, i = 1, 2)$ are both Gaussian integers), then $N(z_1)$ divides $N(z_2)$ in the usual sense of integers.

   (b) Show that for a Gaussian integer z, $N(z) = 1$ if and only if $z = \pm 1, \pm i$, and that these are precisely the z for which z^{-1} is also a Gaussian integer.

   (c) Show that if $z \in \mathbf{Z}[i]$ and $N(z)$ is a prime, then z is not divisible by any Gaussian integer other than $\pm 1, \pm i, \pm z, \pm iz$.

Exercise 3 suggests that one can define primes in $\mathbf{Z}[i]$ in a manner analogous to their definition in $\mathbf{Z}$. One can then show that the unique factorization of a Gaussian integer as a product of primes is true and that there is a way to develop arithmetic using Gaussian integers that has many parallels with the arithmetic of ordinary integers. Gauss was the first mathematician to work systematically with the arithmetical aspects of Gaussian integers. For a nice introduction to these ideas see [HW].

5. Let

$$\omega = \frac{-1 + i\sqrt{3}}{2}$$

Show that

$$\omega^2 = \frac{-1 - i\sqrt{3}}{2} = -\omega - 1$$

and that

$$X^3 - 1 = (X - 1)(X - \omega)(X - \omega^2)$$

Deduce that

$$1, \omega, \omega^2$$

are precisely the three cube roots of 1 in $\mathbf{C}$.

6. In analogy with $\mathbf{Z}[i]$ we can introduce $\mathbf{Z}[\omega]$ consisting of all complex numbers of the form $a + b\omega$ where a, b are integers.

   (a) Show that $a + b\omega = 0$ $(a, b \in \mathbf{Z})$ if and only if $a = b = 0$.

   (b) Verify that

$$(a + b\omega)(a' + b'\omega) = aa' - bb' + (ab' + a'b - bb')\omega$$

It turns out that $[\omega]$ also has the unique factorization property and one can develop, in analogy with $\mathbf{Z}[i]$, the arithmetic of $\mathbf{Z}[\omega]$. The factorization

$$x^3 + y^3 = (x + y)(x + \omega y)(x + \omega^2 y)$$

suggests that the arithmetic of $\mathbf{Z}[\omega]$ may play a role in proving Fermat's last theorem for the exponent 3. See [HW] for a proof along these lines. As we have mentioned earlier, Fermat's last theorem for the exponent 3 was first settled by Euler.

Kummer, **Dedekind** and **Kronecker**, created modern algebraic number theory. The basic and profound difficulty was that the majority of number systems were not like $\mathbf{Q}(i)$ or $\mathbf{Q}(\omega)$ in the sense that one does not have unique factorization of integers in terms of primes in them. The classic example of this is $\mathbf{Q}(\sqrt{-5})$; here the numbers $2, 3, 1 + \sqrt{-5}, 1 - \sqrt{-5}$ are all primes, essentially distinct, and yet we have

$$6 = 2 \times 3 = (1 + \sqrt{-5}) \times (1 - \sqrt{-5})$$

It was only by the dramatic idea of giving up the notion of a prime element and replacing it by the notion of a *prime ideal* that unique factorization in terms of primes was restored. This was a truly revolutionary achievement.

TRIGONOMETRY OF COMPLEX NUMBERS

We shall now discuss another way to calculate with complex numbers. This uses some trigonometry which one would have done in high school. I shall begin by recalling the material that will be useful for us.

The starting point is what are called *trigonometric identities* which are really formulae that express the sine and cosine for the sum of two angles in terms of the sines and cosines of the individual angles. There are two basic formulae given by the following box.

$$\cos(A + B) = \cos A \cos B - \sin A \sin B$$
$$\sin(A + B) = \sin A \cos B + \cos A \sin B$$

We shall now turn to the complex plane and consider a complex number

$$z = a + ib \qquad (a, b \in \mathbf{R})$$

We shall now use *polar coordinates* to represent the point (a, b). We recall that the formulae to go from cartesian to polar coordinates and vice versa are as follows. If R, θ are the polar coordinates of (a, b),

$$a = R \cos \theta$$
$$b = R \sin \theta$$
$$R = \sqrt{a^2 + b^2}$$
$$\theta = \arctan \frac{b}{a}$$

Thus the complex number $z = a + ib$ becomes

$$z = R(\cos \theta + i \sin \theta)$$

The number R, called the r-part, is the distance from the origin; the angle θ, called the angle part, is the angle in the counterclockwise direction that the line joining the point to the origin makes with the real axis. The origin itself is unrepresented; the r-part is 0, but the angle is undetermined. Note also that θ can be changed by adding or subtracting a whole multiple of $360°$ without changing the point. Thus there is an ambiguity in the choice of θ which we must always remember.

What is the use of this new representation? We have seen that in the cartesian representation addition of two complex numbers is easy; one adds the real and imaginary parts separately. But multiplication is complicated. We shall see that in the polar representation multiplication is simple.

Let us therefore take two complex numbers in polar form

$$z = R(\cos A + i \sin A), \qquad u = S(\cos B + i \sin B)$$

and compute their product zu. Using the usual rule for multiplication we see that

$$(\cos A + i \sin A)(\cos B + i \sin B)$$

becomes

$$(\cos A \cos B - \sin A \sin B) + i(\sin A \cos B + \cos A \sin B)$$

which is

$$\cos(A + B) + i \sin(A + B)$$

if we use the identities recalled earlier. Thus we have

$$z = R(\cos A + i \sin A)$$
$$u = S(\cos B + i \sin B$$
$$zu = RS(\cos(A + B) + i \sin(A + B))$$

In other words we have the following extremely simple rule for multiplying in polar coordinates.

Multiply the r-parts of the two numbers separately to get the r-part of the product, and add the angle parts of the two numbers to get the angle part of the product

As a special case consider a complex number where the r-part is 1 so that it is

$$z = \cos A + i \sin A$$

If we multiply any complex number u by z where

$$u = R(\cos B + i \sin B)$$

we get

$$zu = R(\cos(B + A) + i \sin(B + A))$$

In other words, the r-part is unchanged and A is added to the angle part. If you reflect a little this will tell you that the number u is *rotated* by the angle A. Thus we get

Multiplication by $\cos A + i \sin A$

corresponds to rotating by A

This rule also allows us to calculate the powers

$$(\cos A + i \sin A)^N, \quad N = 1, 2, 3, \ldots$$

Each time we multiply by $\cos A + i \sin A$ we add A to the angle part of the number being multiplied. Hence, the result of multiplying 1 by $\cos A + i \sin A$ N times gives us the result we want:

$$(\cos A + i \sin A)^N = \cos NA + i \sin NA \quad (N = 1, 2, 3, \ldots)$$

This formula was first established by **De Moivre** (1667–1754). See the exercises below for an indication of an algebraic proof of this identity. For low values of N one can verify this directly. Thus for $N = 2$ we have

$$(\cos A + i \sin A)^2 = (\cos^2 A - \sin^2 A) + i(2 \cos A \sin A) = \cos 2A + i \sin 2A$$

APPLICATIONS : N^{th} ROOTS

We shall now show how the interpretation of multiplication in the complex plane in polar coordinates allows us to take square roots, cube roots, fourth roots, and in principle, roots of all orders, of any complex number.

First let us determine the polar coordinate representation of some complex numbers. Let

$$z = 1 + i$$

Then

$$R = \sqrt{1 + 1} = \sqrt{2}, \quad \theta = \arctan \frac{1}{1} = 45° = \frac{\pi}{4}$$

Hence

$$1 + i = \sqrt{2}(\cos 45° + i \sin 45°)$$

Again let

$$z = 2\sqrt{3} - 2i$$

Here

$$R = \sqrt{(2\sqrt{3})^2 + (-2)^2} = 4, \quad \theta = \arctan \frac{-1}{\sqrt{3}} = 330°$$

How do we know the angle is 330° and not 150°? This is because the point is $(2\sqrt{3}, -2)$ which is in the *fourth quadrant* and so the angle must be between 270° and 360°. Thus

$$2\sqrt{3} - 2i = 4(\cos 330° + i \sin 330°)$$

Let us now see whether we can find the cube roots of i. We have already done this but the method was just a verification; there was no explanation as to how the cube roots were found in the first place. Let us look for the cube root of i in the form

$$\sqrt[3]{i} = R(\cos A + i \sin A)$$

So we must have

$$[R(\cos A + i \sin A)]^3 = \cos 90° + i \sin 90°$$

This gives

$$R^3 = 1, \qquad (\cos A + i \sin A)^3 = \cos 90° + i \sin 90°$$

Now

$$(\cos A + i \sin A)^3 = \cos 3A + i \sin 3A$$

Since R is positive, $R^3 = 1$ means $R = 1$. So we only have to consider the equation for A. In view of the remark above this comes to

$$\cos 3A = \cos 90°, \quad \sin 3A = \sin 90°$$

What are the solutions A? In geometrical terms, we want to find the angle A such that 3 times the angle is $90°$. Certainly $A = 30°$ is one solution. But there are others. Remember that adding or subtracting a multiple of $360°$ does not change the position of the point in the complex plane. So we should certainly look for A between $0°$ and $360°$. But if we add to A first $120°$ and then $240°$, $3A$ changes by $360°$ and by $720°$ and so will still give the same point represented by $90°$. Moreover there are no other possibilities. In other words, we have exactly three solutions for A:

$$A = 30°, \quad 150°, \quad 270°$$

which give the three cube roots

$$\sqrt[3]{i} = \frac{\sqrt{3} + i}{2}, \quad \frac{-\sqrt{3} + i}{2}, \quad -i$$

confirming our earlier calculation.

This method is obviously perfectly general and will give, for any complex number z (different from zero) and any integer N, N complex numbers which are N^{th} roots of z. Thus if we take

$$z = R(\cos B + i \sin B)$$

and write

$$u = \sqrt[N]{z}$$

Then, taking u in the form

$$u = S(\cos A + i \sin A)$$

we have $u^N = z$ which becomes

$$S^N(\cos A + i\sin A)^N = R(\cos B + i\sin B)$$

As before

$$(\cos A + i\sin A)^N = \cos NA + i\sin NA$$

This gives

$$S^N = R, \qquad \cos NA + i\sin NA = \cos B + i\sin B$$

Since S is positive we must have

$$S = \sqrt[N]{R}$$

which is the unique positive N^{th} of R (Here we must remark that any positive real number has a unique positive N^{th} root). As for A we first have the obvious solution

$$A = \frac{B}{N}$$

But we remember that adding multiples of $\frac{360°}{N}$ to A does not change the position of the point corresponding to NA. Hence we have the N possibilities and no more:

$$A = \frac{B}{N}, \frac{B + 360°}{N}, \frac{B + 720°}{N}, \ldots, \frac{B + (N-1)360°}{N}$$

Hence for u we have the N solutions

$$z = u^N$$
$$u = \sqrt[N]{R}(\cos A + i\sin A),$$
$$A = \frac{B}{N}, \frac{B + 360°}{N}, \frac{B + 720°}{N}, \ldots, \frac{B + (N-1)360°}{N}$$

As another example, let us calculate $\sqrt[5]{i}$. If $t = \sqrt[5]{i}$ we have $t^5 = i$. Taking t in the form $R(\cos A + i\sin A)$ we get

$$R^5(\cos 5A + i\sin 5A) = i = (\cos 90° + i\sin 90°)$$

So as before $R = 1$ and

$$A = 18°, \quad 90°, \quad 162°, \quad 234°, \quad 306°$$

It is known that

$$\sin 18° = \frac{\sqrt{5} - 1}{4}$$

(Can you prove this?). Using $\cos^2 A + \sin^2 A = 1$ we get

$$\cos 18° = \frac{\sqrt{10 + 2\sqrt{5}}}{4}$$

Furthermore, using the formulae

$$\cos 54° = \sin 36° = 2\sin 18° \cos 18°$$

$$\sin 54° = \cos 36° = 1 - 2\sin^2 18°$$

we find

$$\cos 54° = \frac{\sqrt{10 - 2\sqrt{5}}}{4}, \qquad \sin 54° = \frac{\sqrt{5} + 1}{4}$$

Hence the 5 fifth roots of i are

$$\frac{\sqrt{10 + 2\sqrt{5}} + i(\sqrt{5} - 1)}{4}, \; i, \; \frac{-\sqrt{10 + 2\sqrt{5}} + i(\sqrt{5} - 1)}{4}$$

$$\frac{-\sqrt{10 - 2\sqrt{5}} - i(\sqrt{5} + 1)}{4}, \quad \frac{\sqrt{10 - 2\sqrt{5}} - i(\sqrt{5} + 1)}{4}$$

If the angles are randomly chosen we will not have such explicit expressions; we have to use tables for sine and cosine functions. But there is no difficulty in principle.

NOTES AND EXERCISES

1. Prove that for any integer $n \geq 1$, we have

$$(\cos \theta + i \sin \theta)^n = \cos n\theta + i \sin n\theta$$

(Hint : Assume for $n = k$ and prove it for $n = k + 1$ using trigonometric identities for $\cos(u + v)$ and $\sin(u + v)$ in terms of $\cos u, \cos v, \sin u, \sin v$.)

2. Use exercise 1 to prove that

$$\cos n\theta = \cos^n \theta - \frac{n(n-1)}{2} \cos^{n-2} \theta \sin^2 \theta +$$
$$\frac{n(n-1)(n-2)(n-3)}{4} \cos^{n-4} \theta \sin^4 \theta - \ldots$$

Deduce the formulae

$$\cos 4\theta = 8\cos^4 \theta - 8\cos^2 \theta + 1, \qquad \cos 3\theta = 4\cos^3 \theta - 3\cos \theta$$

Also prove by the same method that

$$\sin 3\theta = 3\sin \theta - 4\sin^3 \theta$$

3. Prove that

$$\sin 18° = \frac{\sqrt{5} - 1}{4}, \qquad \cos 18° = \frac{\sqrt{10 + 2\sqrt{5}}}{4}$$

(Hint : It is enough to prove the first formula. Write $\theta = 18°$ and use the identity $\sin 3\theta = \cos 2\theta$ to deduce that $s = \sin\theta$ satisfies $4s^3 - 2s^2 - 3s + 1 = 0$. One root is $s = 1$ which is inadmissible. So s must satisfy $4s^2 + 2s - 1 = 0$ for which the unique positive root is $(\sqrt{5} - 1)/4$.)

4. Prove that the nth roots of 1 are

$$\varepsilon_r = \cos\frac{2\pi r}{n} + i\sin\frac{2\pi r}{n} \qquad (r = 1, 2, \ldots, n)$$

where we write 2π for $360°$ (in radians of course). Hence deduce that if a is a nonzero complex number and z_0 is some n^{th} root of a, then all the nth roots of a are given by

$$z_0\varepsilon_r \qquad (r = 1, 2, \ldots, n)$$

12

FUNDAMENTAL THEOREM OF ALGEBRA

We have mentioned before that **Gauss** was the first mathematician who worked systematically with complex numbers. Gauss is regarded as one of the greatest mathematicians of all time. His work ranges over all of mathematics and natural sciences. He was fully as universal as Euler and even more penetrating.

During his work in algebra he realized that *the problem of constructing explicit formulae for the roots of a given equation was very different from the problem of proving that there are solutions.* Since complex numbers were introduced to compensate for the situation where there were no real solutions to equations over real numbers, he understood that it became essential to prove that once one worked with complex numbers there would always be solutions. The *fundamental theorem of algebra* as proved by Gauss asserts that every equation

$$z^N + a_1 z^{N-1} + a_2 z^{N-2} + \ldots + a_{N-1} z + a_N = 0 \qquad (a_i \in \mathbf{C})$$

has at least one solution in $\mathbf{C}$. Once this is done it is an easy matter to prove that it has N solutions, counting multiplicities. The precise statement of this theorem is as follows.

Theorem *Given any polynomial P with complex coefficients*

$$P(z) \equiv z^N + a_1 z^{N-1} + a_2 z^{N-2} + \ldots + a_{N-1} z + a_N \quad (a_i \in \mathbf{C})$$

there are unique (upto a permutation) complex numbers $r_1, r_2, \ldots, r_N$ (not all distinct) such that

$$P(z) \equiv (z - r_1)(z - r_2) \ldots (z - r_N)$$

The methods Gauss discovered to prove this theorem were truly revolutionary and eventually formed the foundation of the theory of functions of a complex variable. Gauss was so aware of the fundamental importance of this theorem that he constructed many different proofs of it. This was a characteristic habit of Gauss; for those theorems of his which he regarded as fundamental he usually constructed several proofs. Nowadays the fundamental theorem of algebra is proved as one of the applications of the theory of analytic functions of a complex variable; its essence is however a topological argument.

The point of the fundamental theorem is the following. Remember that we introduced complex numbers so that equations with real coefficients would have solutions, if not in the real number system, then at least in the extended system of complex numbers. It is however not to be ruled out that if we now take an equation over the complex numbers, it may not have solutions at all in the complex number system. Amazingly this does not happen. For instance, let us consider a quadratic equation

$$z^2 + az + b = 0$$

The Babylonian method gives the solutions formally as

$$z = \frac{-a \pm \sqrt{a^2 - 4b}}{2}$$

If we now recall that we have proved that every complex number has always two square roots (other than zero which has only one square root, namely zero) *within the complex number system* it follows that the above formula gives two complex roots except when

$$a^2 - 4b = 0$$

when it has only one root, but which is to be viewed as a *double root* in a natural sense. For the cubic equation

$$z^3 + az + b = 0$$

where a, b are now complex numbers, the Cardano formula will give three complex roots, when properly interpreted. The same is true for the biquadratic equation also. Gauss's theorem takes off from this and establishes the amazing result described above.

For polynomials whose coefficients are *real* numbers the fundamental theorem has to be formulated in a different manner. It is easy to see that if z is a root of an equation $P = 0$ where P is a polynomial with real coefficients, then $\bar{z}$, the conjugate of z, is also a root. So the roots of $P = 0$, guaranteed by the fundamental theorem, are either real, or split into pairs of mutually conjugate numbers. If z is not real so that $z \neq \bar{z}$, then

$$(X - z)(X - \bar{z}) = X^2 + aX + b, \qquad a = -(z + \bar{z}), b = z\bar{z}$$

and it is clear that a and b are both real. The fundamental theorem now takes the following form.

Theorem *Let P be a polynomial with real coefficients of degree N. Then*

$$P(z) = (z - r_1)\ldots(z - r_k)(z^2 + 2s_1 z + t_1)\ldots(z^2 + 2s_m z + t_m)$$

where

$$k + 2m = N, \qquad r_i, s_j, t_j \in \mathbf{R}, \qquad s_j^2 < t_j \ \forall j = 1, 2, \ldots, m$$

with the proviso that either k or m can be 0.

There is another point also that is well worth emphasizing. I had remarked earlier that in the process of extending real numbers, which is essentially one of introducing axiomatically the square root of -1 with certain properties, one can start with number systems that are less inclusive than the system of *all* real numbers; in fact we agreed to treat the notion of a general real number somewhat formally. One could be more strict and start with only rational numbers and then adjoin the square root of -1 to get what we called the *Gaussian rationals*, namely numbers of the form

$$a + ib \qquad (a, b \text{ rational numbers})$$

But the fundamental theorem of algebra *will not be true for the system of Gaussian rationals*. For instance, we cannot solve the equation $z^2 = 2$ within the system of Gaussian rationals. In fact, to reach a number system starting with the rationals and for which the fundamental theorem of algebra is valid, and to do this in the most economical manner, one has to make not just the extension by adjoining the square root of -1 but many extensions, even infinitely many of them, adjoining many more roots of equations. A completely explicit description of this minimal extension of the rationals which mathematicians call the *algebraic closure of the rational number system* is still proving elusive. It remains one of the great goals of modern mathematics and it is believed that if and when its solution is discovered, the methods will use everything that mathematicians have discovered so far. Thus the fact that the simple (!) process of adjoining the square root of -1 to the *real number system* leads to a system of numbers for which the fundamental theorem of algebra is true reflects on the deepest structural properties of the real numbers.

CARDANO'S FORMULA FOR

COMPLEX CUBIC EQUATIONS

Now that we have made a little study of complex numbers and some of their properties, it is time to go back and look at Cardano's formula for the solution of a cubic equation. This time we shall be even more general and consider the equation

$$Z^3 + PZ = Q$$

where P, Q are now *complex numbers*! We take the equation in this form, because the elimination of the Z^2 term can be done exactly as before, no change in the procedure being necessary. As usual we shall look for Z as a sum of two cube roots,

$$Z = \sqrt[3]{U} + \sqrt[3]{V}$$

Then

$$Z^3 = U + V + (3\sqrt[3]{U}\sqrt[3]{V})(\sqrt[3]{U} + \sqrt[3]{V})$$

and this expression will be equal to $-PZ + Q$ provided

$$U + V = Q, \qquad \sqrt[3]{U}\sqrt[3]{V} = -\frac{P}{3} \qquad\qquad (*)$$

Thus U and V are roots of the quadratic equation (in T)

$$T^2 - QT - \left(\frac{P}{3}\right)^3 = 0 \qquad (**)$$

If we choose any root, say, U, of this equation, then

$$Z = \sqrt[3]{U} - \frac{P}{3\sqrt[3]{U}}$$

since we must have, by $(*)$,

$$\sqrt[3]{V} = -\frac{P}{3\sqrt[3]{U}}$$

There are three choices for $\sqrt[3]{U}$ and the formula for Z is fixed as soon as one knows the value of $\sqrt[3]{U}$. This gives three values for Z and the solution is complete; in fact, one can show that (see exercise below) if u_1, u_2, u_3 are the three cube roots of U which is a root of $(**)$, and we write

$$z_j = u_j - \frac{P}{3u_j}, \qquad j = 1, 2, 3$$

then one has the factorization

$$Z^3 + PZ - Q = (Z - z_1)(Z - z_2)(Z - z_3)$$

Thus *in the complex domain, Cardano's formula will give all three roots of the cubic equation in question.* One can also verify that the three roots are distinct if and only if $\Delta \neq 0$ (see exercise below).

We are now in a position to dispose of the irreducible case when we considered a cubic equation with real coefficients whose discriminant is < 0. Here there are 3 real roots, but the formula involves complex numbers and it was not clear, when we first encountered this situation, how to interpret the formula. But we shall now see that this poses no problem. We have already shown how the formula of Cardano gives the three roots. All we have to verify now is that these three roots are real. So let us assume that P, Q are real and

$$\Delta = \frac{Q^2}{4} + \frac{P^3}{3} < 0$$

We can therefore write

$$\Delta = -\delta^2$$

where δ is real. Let us write

$$u = \sqrt[3]{U}, \qquad v = -\frac{P}{3u}$$

so that $Z = u + v$. To prove that Z is real we must only show that v is the complex conjugate of u. Now

$$uv = -\frac{P}{3} > 0$$

because

$$\left(\frac{-P}{3}\right)^3 = -\Delta + \frac{Q^2}{4} > 0$$

On the other hand, since u^3 is a root of $T^2 - QT - (P/3)^3 = 0$,

$$u^3 = \frac{Q}{2} \pm \sqrt{\Delta} = \frac{Q}{2} \pm i\delta$$

from which we get

$$|u^3| = |u|^3 = +\sqrt{\frac{Q^2}{4} + \delta^2} = +\sqrt{\frac{Q^2}{4} - \Delta} = \sqrt{\left(\frac{-P}{3}\right)^3}$$

Hence

$$|u^3|^2 = (|u|^2)^3 = \left(\frac{-P}{3}\right)^3$$

showing that

$$|u|^2 = \frac{-P}{3}$$

In other words,

$$u\bar{u} = uv = \frac{-P}{3}$$

from which we conclude at once that

$$v = \bar{u}$$

as we wanted to show.

NOTES AND EXERCISES

1. Verify the factorization of $Z^3 + PZ - Q$ given in the discussion of Cardano's formula in the complex domain. (Hint : Select one cube root u of U and write the cube roots as $u, u\omega, u\omega^2$. Then the three values of Z are

$$u - \frac{P}{3u}, \qquad u\omega - \frac{P}{3u\omega}, \qquad u\omega^2 - \frac{P}{3u\omega^2}$$

   The factorization can now be directly verified using the identities $1 + \omega + \omega^2 = 1 + \omega + 1/\omega = 0$)

2. Prove that the three roots are distinct if and only if $\Delta \neq 0$. (Hint : We have already seen in an exercise to Chapter 8 that if $\Delta = 0$ and $P \neq 0$ (otherwise $P = Q = 0$ and things are trivial), then the roots are $3Q/2P$ (double), $-3Q/P$; the verification was purely algebraic and remains valid when we work over the complex numbers. For proving that if

two roots coincide then $\Delta = 0$, show that $uu' = -P/3$ where u and u' are two distinct cube roots of U. Deduce that $U^2 = (-P/3)^3$. Use the fact that $U^2 - QU - (P/3)^3 = 0$ to conclude that $\Delta = 0$.)

3. Show, using Cardano's formula that the roots of $X^3 - 6X - 4 = 0$ are $-2, 1+\sqrt{3}, 1-\sqrt{3}$.

4. Consider the polynomial algebra but allowing the polynomials to have complex coefficients. Prove the remainder theorem. Deduce that if we know that every polynomial has at least one root, then the factorization as stated in the fundamental theorem will follow at once.

These notes will be seriously incomplete if there is no discussion of **Euler** and **Gauss**. I shall therefore give a very brief presentation of the highlights of the accomplishments of these great mathematicians.

Euler (1707–1783) was the greatest mathematician of his epoch. He was universal and made profound contributions to all parts of mathematics and its applications. His output was prodigious, even after he lost his eyesight late in his life. His Collected Works run to over 100 volumes and are being published by the Swiss Academy of Sciences. He discovered fundamental results in number theory, analysis, elliptic functions, mechanics of solid bodies and fluid mechanics, in graph theory which he founded when he solved the famous problem of the seven bridges of Königsberg, and in topology. We have already mentioned the fact that he found a solution to Fermat's last theorem when the exponent was 3. He found remarkable identities involving infinite series and their sums, such as

$$1 + \frac{1}{2^2} + \frac{1}{3^2} + \ldots = \frac{\pi^2}{6}, \quad 1 + \frac{1}{2^4} + \frac{1}{3^4} + \ldots = \frac{\pi^4}{90}$$

He introduced the idea that divergent series can be summed by different procedures and discovered the famous *Euler-Mclaurin summation formula*. He found the factorization

$$\sum_{n=1}^{\infty} \frac{1}{n^s} = \prod_{p \text{ prime}} \frac{1}{1 - \frac{1}{p^s}}$$

which is at the foundation of all studies of distribution of primes. He found the partial fraction and infinite product expansions of the trigonometric functions,

$$\sin x = x \prod_{n=1}^{\infty} \left(1 - \frac{x^2}{n^2\pi^2}\right), \quad \frac{\pi \cos \pi x}{\sin \pi x} = \frac{1}{x} + \sum_{n=1}^{\infty} \left(\frac{1}{n+x} - \frac{1}{n-x}\right)$$

He also discovered beautiful identities involving the *partition function*, namely the function $p(n)$ that gives, for any positive integer n, the number of ways in which n can be written as a sum of positive integers. He discovered the formula

$$1 + \sum_{n=1}^{\infty} p(n)x^n = \frac{1}{\prod_{n=1}^{\infty}(1 - x^n)}$$

and the very famous identity

$$\prod_{n=1}^{\infty}(1 - x^n) = \sum_{n=-\infty}^{\infty} (-1)^n x^{\frac{1}{2}n(3n+1)}$$

from which he obtained numerous combinatorial results concerning specific types of partitions.

One can go on and on. The reader who wishes to have a deeper understanding of many of Euler's contributions should look into Weil's beautiful account of Euler's life and his mathematics in [W].

Gauss (1777-1855) was without doubt the greatest mathematician of his era which came after Euler's. He was a prodigy and worked in all of mathematics and a great variety of its applications. His work *Disquisitiones Arithmeticae* was a landmark and was the basis for all subsequent research in number theory. He gave a rigorous treatment of the arithmetic of congruences and established the famous theorem of *quadratic reciprocity*. He was the first mathematician to work systematically with complex numbers. He established the fundamental theorem of Algebra. He discovered profound results in the theory of elliptic functions and their applications to number theory. It was his suggestion that led **Riemann** to his famous work on the topology of surfaces and **Betti** to his ideas on the topology of manifolds. In geometry Gauss had obtained many of the fundamental results in *non euclidean geometry* before **Bolyai** and **Lobachevsky** independently discovered the general theory. Gauss's memoir on the geometry of surfaces is regarded as the single most important work in differential geometry until Riemann extended it to manifolds of arbitrary dimensionality in his famous inaugural dissertation in 1854. Finally one of his earliest and yet most beautiful discoveries is the fact that the regular 17–gon can be constructed by straightedge and compass.

In fact, since constructions by straightedge and compass lead to points which are either intersections of a line with a circle or a circle with another circle, the coordinates of such points are obtained by solving a succession of quadratic equations. So the possibility of a Euclidean construction of the regular 17–gon amounts to exhibiting the value of $2\cos\frac{2\pi}{17}$ in terms of successions of square roots. Gauss did this, and his formula is in fact the following:

$$2\cos\frac{2\pi}{17} = \frac{1}{8}\left\{-1+\sqrt{17}+\sqrt{34-2\sqrt{17}}\right\} +$$
$$\frac{1}{8}\sqrt{68+12\sqrt{17}-16\sqrt{34-2\sqrt{17}}-2(1-\sqrt{17})\left(\sqrt{34-2\sqrt{17}}\right)}$$

For a beautiful discussion of this and a very elegant geometric construction of the vertices of the regular 17–gon, see [HW], p. 60.

His work on terrestrial magnetism is recognized by the fact that the fundamental unit of magnetism is called a gauss. He founded the subject of least squares approximation, and the theory of the gaussian distributions, which are basic to all of statistics, probability, and quantum physics. He made many extensive calculations in Astronomy and celestial mechanics. His triangulation of the Hannover region in Germany is a great classic.

The reader who wants to have a more detailed idea of Gauss's mathematics should refer to the book of Gindikin [G], to the translation of Gauss's mathematical diary by J. J. Gray which we have already referred to, as well as a large number of other accounts cited in these references.

13

EQUATIONS OF DEGREE GREATER THAN FOUR

The fundamental theorem of algebra has nothing to say on the problem of finding explicit formulae for solutions of equations. After Ferrari's discovery of the method for solving the biquadratic equation, it became a hot question as to how one can solve equations of degree 5 and higher. However centuries of effort did not yield the solution and people came slowly to the conclusion that perhaps one cannot expect to solve equations of higher degree along the same lines as was possible for equations of degrees 3 and 4. The key point here is how one should interpret the word *along the same lines* in the above remark. People interpreted this to mean that the solution should be in terms of *radicals*, that is, in terms of cube roots, fourth roots, fifth roots, and so on. Then in the nineteenth century this was proved to be impossible. The mathematicians who discovered this were **Abel** (1802–1829), **Galois** (1811–1832), and **Ruffini** (1765–1822). They showed that it is not possible to solve all equations of degree 5 in terms of radicals.

Theorems such as this which assert the impossibility of doing something are inherently very difficult to prove. This is because they assert that there is no way, no matter how clever one tries to be, to do certain things. In fashionable terminology first introduced by physicists, these are *no go theorems*.

One may recall that there have been no go theorems in geometry, such as the theorem of trisecting an arbitrary angle by Euclidean constructions, the problem of constructing a regular polygon with N sides by Euclidean constructions for general values of N, and so on. It is very interesting that the no go theorems of geometrical constructions and the no go theorem of solving algebraic equations by radicals are nowadays viewed from the same perspective, namely that of the *theory of groups*. This is the extraordinary legacy of Galois.

The biggest no go theorem of all involves the structure of mathematics itself. In the beginning of the twentieth century **Hilbert** examined the question as to whether every statement that is made in arithmetic can be resolved to be either true or false. He believed that this was always possible. But to the great surprise of Hilbert as well as all the mathematicians, **Gödel** (1906–1978) proved in the 1930's that this is too optimistic and that in any sufficiently complicated mathematical theory there are statements that are true but cannot be demonstrated to be so. These are however matters that can be pursued only after a substantial background in mathematics has been acquired by the aspiring student.

14

GENERAL NUMBER SYSTEMS AND THE

AXIOMATIC TREATMENT OF ALGEBRA

The introduction of complex numbers was so stunning an event that it was natural that people tried to understand everything about the process and in particular why it was so successful. One of the most striking things about the complex numbers was their geometric interpretation. Especially striking is the fact that we have talked about at length, namely the interpretation of multiplication in terms of rotations in the plane. In the nineteenth century when big developments in mathematics and physics were taking place, it became an intensely debated question whether one could discover other number systems which would model rotations in three dimensional space. Many of course tried this problem but it was given only to **Hamilton** (1805–1865) to come up with the solution. He invented what are now called *quaternions* for this purpose. The system of quaternions is universally denoted by **H** in honour of Hamilton.

Before understanding the nature and scope of Hamilton's discovery let us informally comment on what one means by a *number system*. Taking into account what we have discussed so far we may say that a number system is a set of objects that are represented by ordered N-tuples of real numbers

$$\mathbf{a} = (a_1, a_2, \ldots, a_N)$$

For example, complex numbers fit this description as we can take $N = 2$. These are added in the obvious way:

$$\mathbf{a} + \mathbf{b} = (a_1 + b_1, \ldots a_N + b_N)$$

if

$$\mathbf{a} = (a_1, \ldots, a_N), \quad \mathbf{b} = (b_1, \ldots, b_N)$$

We should also define multiplication. The example of complex numbers suggests that the rule for multiplication could be very involved if interesting examples are to be constructed. The best way to proceed is to say that there are N^3 real numbers, called *structure constants*, denoted by

$$A_{ijk} \qquad (i, j, k = 1, 2, \ldots, N)$$

which define multiplication:

$$(a_1, \ldots, a_N) \times (b_1, \ldots, b_N) = (c_1, \ldots, c_N)$$

where

$$c_i = \sum_{jk} A_{ijk} a_j b_k \qquad (i = 1, 2, \ldots, N)$$

A system such as this is called an *algebra*. The element whose components are all 0 is called the *zero element, or simply, zero* of the algebra. If we write

$$\mathbf{a} = (a_1, a_2, \ldots, a_N)$$

and the multiplication has the property that

$$\mathbf{a} \times (\mathbf{b} \times \mathbf{c}) = (\mathbf{a} \times \mathbf{b}) \times \mathbf{c}$$

for all $\mathbf{a}, \mathbf{b}, \mathbf{c}$, then we say that we have *an associative algebra*. Usually we require also that there is a unit number, i.e., an element $\mathbf{e}$ such that

$$\mathbf{e} \times \mathbf{a} = \mathbf{a} \times \mathbf{e} = \mathbf{a}$$

We shall from now on drop the sign $\times$ and write

$$\mathbf{a} \times \mathbf{b} = \mathbf{ab}$$

Among these the number systems are singled out by the following crucial property:

Every element $\mathbf{a}$ *which is not zero has an inverse in the sense of the multiplication and unit defined above, i.e., there is an element written* $\mathbf{a}^{-1}$ *such that*

$$\mathbf{aa}^{-1} = \mathbf{a}^{-1}\mathbf{a} = \mathbf{e}$$

An algebra with this property is called a *division algebra*.

Often the multiplication law is such that

$$\mathbf{ab} = \mathbf{ba}$$

for all $\mathbf{a}, \mathbf{b}$. Then the algebra is said to be *abelian* or *commutative*. Commutative division algebras are called *fields*. As examples of associative and commutative algebras that are not fields one can mention the polynomial algebra of all polynomials

$$P(X) = a_0 X^n + a_1 X^{n-1} + \ldots + a_n \qquad (a_i \in \mathbf{R} \ \forall i)$$

Hamilton realized that if a number system is to model rotations in three dimensional space, then multiplication has to be *noncommutative*. This means, as we have explained before, that for two numbers $\mathbf{a}$ and $\mathbf{b}$, $\mathbf{ab}$ is not always equal to $\mathbf{ba}$. In some mysterious way motivated by rotations about the three coordinate axes in three dimensional space Hamilton introduced not one, or two, but *three* imaginary quantities

$$\mathbf{i}, \quad \mathbf{j}, \quad \mathbf{k}$$

satisfying the following rules:

$$\mathbf{i}^2 = \mathbf{j}^2 = \mathbf{k}^2 = -1$$
$$\mathbf{ij} = -\mathbf{ji} = \mathbf{k}$$
$$\mathbf{jk} = -\mathbf{kj} = \mathbf{i}$$
$$\mathbf{ki} = -\mathbf{ik} = \mathbf{j}$$

The *quaternion algebra* is then the system $\mathbf{H}$ of objects represented as

$$\mathbf{q} = a + b\mathbf{i} + c\mathbf{j} + d\mathbf{k} \qquad (a, b, c, d \in \mathbf{R})$$

and rules of addition and multiplication are as expected, with $\mathbf{i}, \mathbf{j}, \mathbf{k}$ obeying the rules set forth as above. It is then astonishing to discover that this is a number system, except that multiplication depends on the order of the numbers that are being multiplied. However $\mathbf{H}$ is associative as can be verified. For the inverse of a nonzero quaternion we have the formula

$$(a + b\mathbf{i} + c\mathbf{j} + d\mathbf{k})^{-1} = (a^2 + b^2 + c^2 + d^2)^{-1}(a - b\mathbf{i} - c\mathbf{j} - d\mathbf{k})$$

Hamilton showed that the quaternions can be used to model rotations in three dimensional space. Decades later, when quantum mechanics was invented, the physicists (such as **Dirac** (1902–1984)) found that quaternions were indispensable in understanding the *quantum spin of the electron.*

The story of the number systems does not end here. The discovery of Hamilton heralded the golden age of algebra when whole families of algebraic systems were discovered. Mathematicians used to call them *hypercomplex systems*; nowadays they are called *algebras.* They were found to be useful in an extraordinary number of problems, both in mathematics and physics. For example, there is a system called the *Clifford algebra* invented by the English mathematician **Clifford** (1845–1879) which plays the same role in the study of rotations in N-dimensional space as the quaternions do in three dimensional space. *However, none of these are division algebras; they do not have the property that nonzero elements are invertible.* The problem of number systems, in the formulation I have given, has only three solutions:

$$\mathbf{R}, \quad \mathbf{C}, \quad \mathbf{H}$$

if one insists on the associativity of multiplication. This was proved by **Frobenius** (1848–1917) in the latter part of the nineteenth century. If one is allowed to drop associativity, then there are additional solutions. There is an 8-dimensional system called *Cayley numbers* which was first discovered by the English mathematician **Cayley** (1821–1895). It is also called *Octonions* because they use 7 imaginary quantities! Although not associative, it satisfies a weaker version of the associativity law. That there are no division algebras, associative or not, other than in dimensions 1, 2, 4, 8 was proved by topological methods in the 1960's. No algebraic proof of this result is known.

The work of nineteenth century algebraists on general types of algebras remained largely unnoticed outside the world of mathematics until the problems of quantum physics proved to be beyond the resources of classical mathematics to solve. It was then discovered that these algebraic systems contained the key ingredients for the formulation and solution of the fundamental questions of quantum theory. One of the most famous and spectacular instances of this is Dirac's discovery of the equations of motion of an electron that obeys both the principles of quantum theory and the theory of special relativity. Dirac's equation depends in an essential way on the structure and properties of the Clifford algebra mentioned above. The axiomatic theory of algebras should therefore not be dismissed as the product of the disordered imaginations of mathematicians. They are indispensable for solving a whole range of problems in the physical sciences which are at the heart of the astounding progress in technology that is being made today.

NOTES AND EXERCISES

1. Let us identify the quaternion $a_0 + a_1\mathbf{i} + a_2\mathbf{j} + a_3\mathbf{k}$ with $\mathbf{a} = (a_0, a_1, a_2, a_3)$. Write

$$\mathbf{ab} = \mathbf{c}$$

Calculate the formulae for the c_j in terms of the a_k, b_ℓ.

In the definition of an algebra we assumed that the elements can be represented by n-tuples of real numbers. If we replace real numbers by rational numbers in the definitions we widen the concept of algebra. The next two exercises examine some examples of this generalization.

2. Let $\mathbf{Q}(\sqrt{2}, i)$ be the set of complex numbers obtained by performing all algebraic operations on $\sqrt{2}$ and i; this is called the *field generated by* $\sqrt{2}, i$. Show that this is the set of all numbers of the form

$$z = a + bi + c\sqrt{2} + di\sqrt{2} \qquad (a, b, c, d \in \mathbf{Q})$$

Similarly describe the field $\mathbf{Q}(\sqrt{2}, \sqrt{3})$. In both cases find the formula for the inverse of a nonzero element.

3. Let $\mathbf{Q}_n$ be the field generated by all the nth roots of 1. This is called the nth *cyclotomic field*, the word coming from the fact that these roots divide the unit circle in the complex plane into n equal parts. Show that $\mathbf{Q}_2 = \mathbf{Q}(i)$ and $\mathbf{Q}_3 = \mathbf{Q}(\omega)$. Show also that the elements of $\mathbf{Q}_5$ are precisely the numbers of the form

$$z = r + s\theta + t\theta^2 + u\theta^3 \qquad (r,s,t,u \in \mathbf{Q})$$

where

$$\theta = \cos 72° + i \sin 72°$$

and

$$1 + \theta + \theta^2 + \theta^3 + \theta^4 = 0$$

The cyclotomic fields play a major role in number theory. Among their earliest appearances is in the work of Gauss on the construction of the regular 17–gon by straightedge and compass. His solution was based on a deep analysis of the cyclotomic field $\mathbf{Q}_{17}$. For a detailed discussion of this question the reader may refer to [G] [HW].

Matrices. It is not possible to construct a reasonably interesting theory of algebras unless one has a large supply of concrete algebras. The simplest of these are the so-called *matrix algebras*.

A *matrix of order* n is a square array of n rows and n columns with each entry a number, real or complex. If the entries are real, it is called a *real matrix*, otherwise it is called a *complex matrix*. For instance

$$\begin{pmatrix} 1 & 2 \\ 2 & 1 \end{pmatrix}$$

is a real matrix of order 2. One writes

$$\begin{pmatrix} a_{11} & a_{12} & \cdots & a_{1n} \\ a_{21} & a_{22} & \cdots & a_{2n} \\ \vdots & \vdots & \ddots & \vdots \\ a_{n1} & a_{n2} & \cdots & a_{nn} \end{pmatrix}$$

where the a_{ij} are numbers, real or complex. We write this in the abbreviated form as

$$(a_{ij})_{1 \le i,j \le n}$$

The operations of addition and subtraction are defined by simply adding or subtracting corresponding entries. Thus

$$(a_{ij}) \pm (b_{ij}) = (c_{ij}), \qquad c_{ij} = a_{ij} \pm b_{ij}$$

Multiplication is however a little more involved. If A, B are matrices of order n, then the product $AB = C$ is defined as the matrix (c_{ij}) where

$$C = (c_{ij}), \qquad c_{ij} = \sum_{k=1}^{n} a_{ik} b_{kj}$$

In other words, the ij–th element of C is obtained by multiplying the elements of the i-th row of A by the corresponding elements of the j–th column of B and adding up the results. This gives us an algebra which is associative, but not commutative when $n \geq 2$, and has a unit element I_n, which is the matrix

$$I_n = \begin{pmatrix} 1 & 0 & \cdots & 0 \\ 0 & 1 & \cdots & 0 \\ \vdots & \vdots & \ddots & \vdots \\ 0 & 0 & \cdots & 1 \end{pmatrix}$$

We can identify the algebra of complex numbers with an algebra of matrices by making the correspondence

$$a + ib \longrightarrow \begin{pmatrix} a & b \\ -b & a \end{pmatrix} \qquad (a, b \in \mathbf{R})$$

It is easy to verify that this correspondence preserves addition, subtraction, and multiplication, takes 1 to I_2, and 0 to 0, the matrix all of whose entries are 0. We can also construct a correspondence of $\mathbf{H}$ with the algebra of matrices of order 2. Let us define the matrices $\sigma_j, j = 1, 2, 3$ by

$$\sigma_1 = \begin{pmatrix} 0 & i \\ i & 0 \end{pmatrix}, \qquad \sigma_2 = \begin{pmatrix} 0 & 1 \\ -1 & 0 \end{pmatrix}, \qquad \sigma_3 = \begin{pmatrix} -i & 0 \\ 0 & i \end{pmatrix},$$

A simple calculation shows that

$$\sigma_1^2 = \sigma_2^2 = \sigma_3^2 = -I_2$$

and

$$\sigma_1\sigma_2 = -\sigma_2\sigma_1 = \sigma_3, \quad \sigma_2\sigma_3- = -\sigma_3\sigma_2 = \sigma_1, \quad \sigma_3\sigma_1 = -\sigma_1\sigma_3 = \sigma_2$$

These are precisely the relations satisfied by the quaternions $\mathbf{i}, \mathbf{j}, \mathbf{k}$. So the correspondence

$$a_0 + a_1\mathbf{i} + a_2\mathbf{j} + a_3\mathbf{k} \longrightarrow a_0 I_2 + a_1\sigma_1 + a_2\sigma_2 + a_3\sigma_3 \quad (a_r \in \mathbf{R})$$

will map the quaternion algebra inside the algebra of complex matrices of order 2 preserving all the operations. This becomes, after a routine calculation, the correspondence

$$a_0 + a_1\mathbf{i} + a_2\mathbf{j} + a_3\mathbf{k} \longrightarrow \begin{pmatrix} a_0 - ia_3 & a_2 + ia_1 \\ -a_2 + ia_1 & a_0 + ia_3 \end{pmatrix}$$

These two examples clearly indicate that the algebras of matrices of various orders are very versatile tools in algebra.

Quaternions and rotations. We mentioned that the problem of modeling rotations in three dimensional space by elements of a number system was a prime motivation in the discovery of the quaternion algebra. We shall now indicate how this can be done. Let us recall that in the complex plane, rotations are modeled by multiplication by the complex numbers of the form

$$\cos\theta + i\sin\theta$$

which are precisely the complex numbers of absolute value 1. There is an analogous construction in the quaternion algebra. For any quaternion

$$u = a_0 + a_1\mathbf{i} + a_2\mathbf{j} + a_3\mathbf{k}$$

define its *conjugate* as the quaternion

$$u^* = a_0 - a_1\mathbf{i} - a_2\mathbf{j} - a_3\mathbf{k}$$

Then it is easy to check that

$$uu^* = u^*u = a_0^2 + a_1^2 + a_2^2 + a_3^2$$

so that

$$uu^* = 0 \Longleftrightarrow u^*u = 0 \Longleftrightarrow u = 0$$

and for $u \neq 0$ its inverse is given by

$$u^{-1} = \frac{1}{uu^*}u^*$$

Quaternionic conjugation is very similar to complex conjugation. It preserves sum and differences, and

$$(u^*)^* = u$$

There is a difference though; under multiplication, it reverses the order of multiplication. More precisely,

$$(u_1u_2)^* = u_2^*u_1^*$$

Finally, the real numbers are identified as the quaternions whose coefficients of $\mathbf{i}, \mathbf{j}, \mathbf{k}$ are zero, and these commute with all quaternions.

Let us now introduce two special sets of quaternions. The set of all *unit quaternions* is denoted by U. It is the set defined by

$$U = \{x \in \mathbf{H} \mid xx^* = 1\}$$

It is clear that all elements of U are invertible, that $1 \in U$, and that if $x_1, x_2 \in U$, then $x_1 x_2^{-1} \in U$. Moreover,

$$x^{-1} = x^* \qquad (x \in U)$$

We also denote by $\mathbf{H}_0$ the set of all quaternions for which the real part is 0. In fact, for any quaternion

$$u = a_0 + a_1\mathbf{i} + a_2\mathbf{j} + a_3\mathbf{k}$$

we call a_0 *the real part* of u, written $\Re(u)$. Then we have,

$$2a_0 = u + u^*$$

exactly as in the case of the algebra of complex numbers. We now claim that

$$x \in U, u \in \mathbf{H}_0 \Longrightarrow xux^{-1} \in \mathbf{H}_0$$

To see this, note that for any $x \in U$ and $u \in \mathbf{H}$,

$$
\begin{aligned}
2\Re(xux^{-1}) = 2\Re(xux^*) &= (xux^* + (xux^*)^*) \\
&= (xux^* + xu^*x^*) = x(u + u^*)x^* \\
&= x(2\Re(u))x^* = 2\Re(u)xx^* = 2\Re(u)
\end{aligned}
$$

so that

$$x \in U, u \in \mathbf{H}_0 \implies xux^{-1} \in \mathbf{H}_0$$

We now identify the set $\mathbf{H}_0$ with the three dimensional space by the correspondence

$$u = a_1\mathbf{i} + a_2\mathbf{j} + a_3\mathbf{k} \longrightarrow (a_1, a_2, a_3)$$

Then, for any unit quaternion x, namely any element of U, the transformation

$$u \longrightarrow xux^{-1} \qquad (u \in \mathbf{H}_0)$$

goes over, under the above correspondence, to a transformation

$$R_x : (a_1, a_2, a_3) \longrightarrow (b_1, b_2, b_3)$$

One can show that this is a rotation in three dimensional space. In fact, for any quaternion

$$v = c_0 + c_1\mathbf{i} + c_2\mathbf{j} + c_3\mathbf{k}$$

we have

$$c_0^2 + c_1^2 + c_2^2 + c_3^2 = vv^*$$

and so, since

$$uu^* = (xux^{-1})(xux^{-1})^*$$

we see that

$$a_1^2 + a_2^2 + a_3^2 = b_1^2 + b_2^2 + b_3^2$$

showing that R_x preserves the distance in the three dimensional space. It can be shown that R_x is a rotation, i.e., it preserves the orientation of the space, and that all rotations in the three dimensional space may be obtained in this manner. The proofs of these statements are a little more difficult.

Clifford algebra and Dirac's equation for the electron. We shall now briefly touch upon one of the most spectacular applications of matrix algebras, namely, the derivation of the equations of motion of the spinning electron in quantum mechanics, first obtained by the great physicist **Dirac**. When Dirac started investigating the equations of motion of a relativistic electron, the only equation available was the so–called *Klein–Gordon* equation which was

$$\psi_{00} - \psi_{11} - \psi_{22} - \psi_{33} + m^2\psi = 0 \qquad (\psi_{rr} = \partial^2\psi/\partial x_r^2)$$

where x_0, x_1, x_2, x_3 are the coordinates of a point in spacetime, ψ is the wave function of the electron of mass m. Dirac's point of departure was his observation that this equation has to be replaced by another one in which only the first derivatives of ψ entered. So he tried to construct a first order differential operator whose square is the operator above, that is, to find

$$L = \gamma_0\partial/\partial x_0 + \gamma_1\partial/\partial x_1 + \gamma_2\partial/\partial x_2 + \gamma_3\partial/\partial x_3$$

so that

$$L^2 = \partial^2/\partial x_0^2 - \partial^2/\partial x_1^2 - \partial^2/\partial x_2^2 - \partial^2/\partial x_3^2$$

A simple calculation shows that the coefficients γ_j must satisfy

$$\gamma_0^2 = 1, \gamma_j^2 = -1 \ (j = 1, 2, 3)$$
$$\gamma_j \gamma_k + \gamma_k \gamma_j = 0 \ (j \neq k, j, k = 0, 1, 2, 3) \qquad (*)$$

If we had assumed that the γ_j are complex numbers, then at least one of them would have to be 0 which contradicts the requirement that their squares are ± 1. So Dirac abandoned this assumption and looked for *matrices* that satisfied these properties. Miraculously he found that such systems of matrices exist, that they have to be of order a multiple of 4, and that one can find a solution of order 4. Let us recall the matrices σ_j of order 2 defined in connection with the quaternion algebra. Then Dirac found that the matrices of order 4 defined by

$$\gamma_0 = \begin{pmatrix} 0 & I_2 \\ I_2 & 0 \end{pmatrix}, \qquad \gamma_j = \begin{pmatrix} 0 & i\sigma_j \\ -i\sigma_j & 0 \end{pmatrix}$$

satisfy the equations $(*)$. So Dirac's equation becomes

$$\gamma_0 \partial \psi / \partial x_0 + \gamma_1 \partial \psi / \partial x_1 + \gamma_2 \partial \psi / \partial x_2 + \gamma_3 \partial \psi / \partial x_3 = \pm im\psi$$

The fact that the γ_j are 4×4 matrices necessarily implies that ψ must be a vector with 4 components and therefore the spin of the electron is automatically included in this model ! Nowadays we think of the variables γ_j as generators of an algebra in which we use the relations $(*)$ while performing any algebraic operation with the γs. Such algebras (in any number of the γ variables) are called *Clifford algebras* or *Clifford numbers* in honour of **Clifford** who first introduced them. This is one more instance when concepts worked out many decades earlier, become the essential tool in fundamental physical applications that ceme later.

Every Clifford number has a unique representation in the form

$$x = a_0 + \sum_i a_i \gamma_i + \sum_{i<j} a_{ij} \gamma_i \gamma_j + \sum_{i<j<k} a_{ijk} \gamma_i \gamma_j \gamma_k + a_{0123} \gamma_0 \gamma_1 \gamma_2 \gamma_3 \qquad (Cl_4)$$

because any product of the γs can be reduced to one of the products (upto a sign)

$$\gamma_i, \gamma_i \gamma_j \ (i < j), \gamma_i \gamma_j \gamma_k \ (i < j < k), \gamma_0 \gamma_1 \gamma_2 \gamma_3$$

using the rules $(*)$. Addition and subtraction of Clifford numbers is by addition and subtraction of the coefficients. For multiplication one has to use $(*)$. For example

$$\gamma_0 \gamma_2 \gamma_3 \gamma_1 \gamma_2 = \gamma_0 \gamma_1 \gamma_3, \qquad \gamma_0 \gamma_2 \gamma_3 \gamma_2 \gamma_3 = -\gamma_0$$

If we compare (Cl_4) with the quaternion algebra in which the elements have the form

$$x = a_0 + a_1 \sigma_1 + a_2 \sigma_2 + a_3 \sigma_1 \sigma_2, \qquad \sigma_1^2 = \sigma_2^2 = -1, \ \sigma_1 \sigma_2 + \sigma_2 \sigma_1 = 0$$

we see that Clifford algebras are a vast generalization of the algebra of quaternions; they retain the associativity property, but not the division algebra property.

To go on further and indicate the many applications of Clifford numbers to algebra, geometry, analysis, and physics, will require a very substantial preparation and we shall not do so here. We hope that these notes have shown the reader a glimpse of the vast applications that are within the scope of the modern theory of algebras.

References

[DS] B. Dutta and A. N. Singh, *History of Hindu mathematics, I and II*, Asia Publishing House, 1962.

[E] H. Eves, *An introduction to the history of mathematics*, Holt, Rinehart, and Winston, 1976.

[G] S. G. Gindikin, *Tales of physicists and mathematicians*, Birkhäuser, 1988.

[H–A] T. L. Heath, *The works of Archimedes*, Dover.

[H–E] T. L. Heath, *Euclid's elements*, Vols I, II, III (Books 1–13), Dover.

[HW] G. H. Hardy, and E. M. Wright, *An introduction to the theory of numbers*, Oxford, 1960.

[RW] Gerolamo Cardano : *Ars Magna : The great art or the rules of Algebra*, Translated and edited by T. Richard Witmer, M. I. T. Press, 1968.

[S] H. Stark, *An introduction to number theory*, M. I. T. Press, 1978.

[vW1] B. L. van der Waerden, *Geometry and Algebra in ancient civilizations*, Springer-Verlag, 1983.

[vW2] B. L. van der Waerden, *A history of Algebra*, Springer-Verlag, 1985.

[W] A. Weil, *Number theory : an approach through history from Hammurapi to Legendre*, Birkhäuser, 1984.

Chronology

A
Abel (1802–1829)
Al–Khwarizmi (c. 780–c. 850)
Apollonius (c. 262 BC–190 BC)
Archimedes (c. 287 BC–212 BC)
Āryabhaṭa (476–c. 550)
B
Beltrami (1835–1900)
Bhāskara (c. 1114–1185)
Bolyai (1802–1860)
Bombelli (1526–1573)
Brahmagupta (c. 598–665)
C
Cantor (1845–1918)
Cardano (1501–1576)
Cayley ((1821–1895)
Clifford (1845–1879)
D
De Moivre (1667–1754)
Dedekind (1831–1916)
Diophantus (c. 250)
Dirac (1902–1984)
E
Euclid (c. 300 BC)
Eudoxus (c. 400 BC)
Euler (1707–1783)
F
Fermat (1601–1665)
Ferrari (1522–1565)
Ferro (Scipione del) (1465–1526)
Fibonacci (1180–1240)
Frobenius (1848–1917)
G
Galois (1811–1832)
Gauss (1777–1855)
Gödel (1906–1978)
Gregory (1638–1675)

H
Hamilton (1805–1865)
Hermite (1822–1901)
Hilbert (1862–1943)
J
Jayadeva (c. 1100)
K
Klein (1849–1925)
Kronecker (1823–1891)
Kummer (1810–1893)
L
Lagrange (1736–1813)
Leibniz (1646–1716)
Lindemann (1852–1939)
Liu Hui (c. 300)
Lobachevsky (1793–1856)
N
Newton (1642–1727)
O
Omar Khayyam (c. 1150)
P
Pacioli (1445-1514)
Pasch (1843–1930)
Pythagoras (c. 500 BC)
R
Ramanujan (1887–1920)
Riemann (1826–1866)
Ruffini (1765–1822)
T
Tartaglia (c. 1500–1557)
Tabit ben Qurra (836–931)
V
Viete (1540–1603)
W
Weil (1906–)
Weyl (1885–1955)

Index